Electronic tes[illegible] equipment ha[illegible]book

Electronic test equipment handbook

Steve Money

PC Publishing

PC Publishing
4 Brook Street
Tonbridge
Kent TN9 2PJ

First published 1990

ISBN 1 870775 11 2

British Library Cataloguing in Publication Data

Money, Steve A.
Electronic test equipment handbook.
1. Electronic testing equipment
I. Title
621.381548

ISBN 1-870775-11-2

Phototypesetting by Scribe Design, Gillingham, Kent
Printed and bound by BPCC Wheatons, Exeter

Preface

Electronics engineers and technicians have available to them a wide selection of test and measurement instruments. These range in complexity from simple continuity testers, through multimeters and oscilloscopes, to complex instruments such as logic analysers. In this book we shall take a look at the principles of operation of the various types of instrument and see how they may be used for a selection of measurement and test applications.

Analogue readout meters are still widely used for voltage and current measurement, and the principles of operation and application of these meters are described in Chapter 1. In recent years digital multimeters have become popular and are now widely used by electronics engineers. Chapter 2 examines these instruments which have the advantage of a precise digital readout and usually incorporate features such as automatic polarity indication and automatic range selection.

Various analogue methods of measuring component values, such as resistance, inductance and capacitance, are dealt with in Chapter 3 and include techniques using analogue meters and bridge circuits. Modern component testers often use digital techniques, and a brief description of these is given.

The oscilloscope is perhaps next in popularity after the multimeter as a test instrument and is vital for examination of signal waveforms. The basic principles and features of oscilloscopes are described together with some typical applications. Many modern laboratory oscilloscopes provide digital storage facilities and the basic principles of these instruments are described.

For experimental work and for alignment of equipment, various types of signal sources are available ranging from simple sine wave oscillators to digitally controlled signal generators. Some modern instruments provide sine, square and triangular waveform outputs and a typical circuit for this type of instrument is described.

Among the more complex instruments available for laboratory use are the digital logic analysers and various forms of spectrum analyser. The basic principles of these instruments are discussed in Chapter 7. Many of the modern laboratory instruments include facilities for remote control and for data transfers to and from a computer. This is usually implemented by using the IEEE488 General Purpose Interface Bus system which is described in the final chapter of the book.

Contents

1 Analogue meters

Of all the test instruments available to the electronics engineer, perhaps the most widely used is the analogue type meter which is used for measuring voltage or current. In this type of meter a pointer moves over a calibrated scale to indicate the level of the voltage or current being measured. Meters of this type are often used as panel meters for displaying voltage or current levels in a piece of equipment. Simple meters of this type may be used when testing a piece of equipment and for experimental work where the monitoring of voltage or current levels at various points in the circuit under test is required. For general diagnostic work a multi-purpose meter, such as the Avometer Model 8, provides a more flexible tool. Unlike the panel meter, which has a fixed function and sensitivity, the multimeter provides a number of switched ranges for measurement of both current and voltage levels and often includes other functions such as the measurement of resistance.

A number of different techniques may be used to produce the pointer movement in an analogue meter. Two types which are likely to be encountered by the electronics engineer are the moving iron and moving coil meters in which the mechanical force used to move the pointer is produced by the interaction of two magnetic fields. Another type of meter which may occasionally be seen is the electrostatic voltmeter which uses the force generated when a high voltage is applied between two plates to produce an electrostatic field.

Moving iron meter

The simplest type of analogue meter in popular use is the moving iron meter which has the basic construction shown in Figure 1.1.

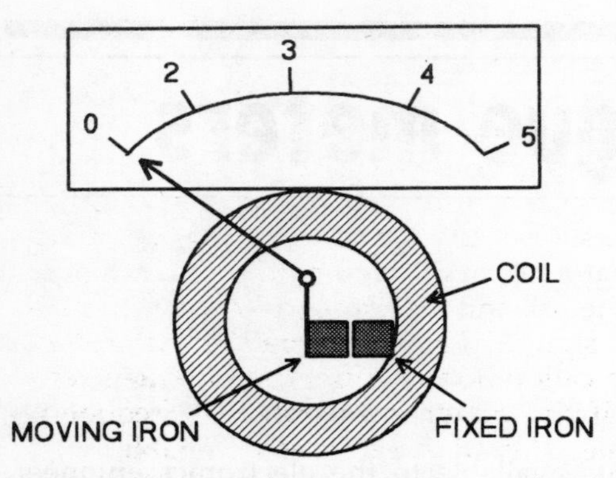

Fig 1.1 Construction of a moving iron meter.

The current to be measured flows through a fixed coil and produces a magnetic field which magnetises two elements made of soft iron. One element is fixed to the coil assembly whilst the other is free to move and is linked to the bottom end of a pointer.

When a current passes through the coil it produces a magnetic field along the axis of the coil. The magnetic field passes through the two pieces of iron which become magnetised so that the N magnetic poles on the two iron elements are adjacent to one another as also are the two S poles. Since like magnetic poles repel one another the result is that a mechanical force is developed between the iron elements which tries to move them apart. The piece of iron attached to the coil is unable to move but the movable one is pushed away from the the fixed element and moves the pointer. The pointer is pivoted a short distance away from the point where the iron element is attached to it so that although the piece of iron only moves a short distance the opposite end of the pointer moves through a large arc over the calibrated scale. The amount by which the pointer moves depends upon the level of current flowing through the coil.

In a moving iron meter the mechanical force produced is roughly proportional to the square of the current flowing in the coil. The result is that the scale follows a square law with divisions close together at the low current end of the scale and spread out at the high current end. As the two pieces of iron move further apart the interaction between their magnetic fields tend to get weaker so the scale readings near full scale are slightly closer together than those at the centre. On a typical meter with a scale of 0 to 5 the scale is so cramped at the lower end that the markings between 0 and 2 are not shown and the 2 position is very close to the 0 position.

The moving iron meter has the advantage that it has a relatively simple construction and is therefore cheap to manufacture. The meter is also quite robust so that it is well suited to applications where it may be subject to shocks or vibration. Another advantage of this type of meter is that it is not affected by the polarity of the current flowing through the coil and therefore will respond equally well to both DC and AC signals. For AC operation the meter is designed so that its scale calibration is correct for frequencies in the range 45 to 65 Hz. If the meter is operated at frequencies outside this range the scale calibration becomes unreliable.

The main disadvantages of the moving iron meter are that it has a non linear scale and is relatively insensitive with a typical voltmeter taking as much as 10mA of current at full scale. Some types of moving iron meter use specially shaped iron elements to produce a more linear scale calibration. The moving iron meter is also sensitive to external magnetic fields so it is important to ensure that meters of this type are located well away from permanent magnets or conductors carrying heavy currents.

This type of meter is generally used as a panel mounted indicator to measure heavy currents, such as in a battery charger, or power supply voltages where its poor sensitivity becomes unimportant and its ability to respond to AC voltages is an advantage. It is also widely used in automobile applications.

Moving coil meter

The most widely used type of analogue meter is the moving coil type which has the general arrangement shown in Figure 1.2. A permanent magnet is used to provide a strong magnetic field between a pair of shaped pole pieces. If a coil of wire is placed between the poles and electric current is passed through the coil, a magnetic field is produced along the axis of the coil. This magnetic field interacts with the field from the permanent magnet to produce a mechanical force between the coil and the magnet. When the coil is mounted on a spindle the effect of the interaction of the magnetic fields is that the coil rotates relative to the fixed magnet. The principle is similar to that involved in an electric motor except that in the meter the rotation is limited to an angle of about 90 degrees.

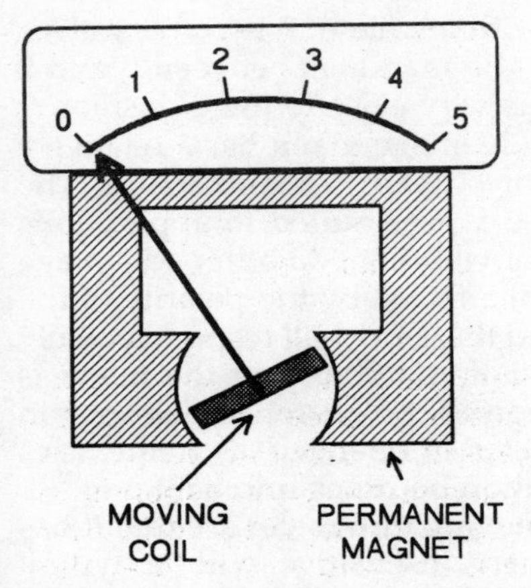

Fig 1.2 Construction of a moving coil meter.

The current is fed to and from the moving coil via a pair of flat coiled springs similar to the hairspring in a clock. These springs also provide a restraining torque and damp down the movement of the coil so that it does not oscillate. The shaft on which the coil is mounted has tapered ends which fit into jewelled bearings to produce minimum friction. A pointer connected to the coil is used to indicate the degree of rotation of the coil and the remote end of this pointer moves over a calibrated scale to give a reading for the current flowing in the coil.

For a moving coil meter the angle of rotation of the coil is directly proportional to the current flowing through it so that the scale calibration is linear. The sensitivity of the meter is determined by the strength of the field from the permanent magnet and the number of turns of wire on the coil.

In a typical moving coil meter the rotation of the coil is usually limited to about 90 degrees but some types of meter are designed to allow the pointer to move over an angle of up to 270 degrees to give an expanded scale without increasing the physical size of the meter unit.

The main advantages of the moving coil type meter are its high sensitivity and its linear scale. Meters which produce full scale readings for a current of only 100 microamperes are common and are widely used as the basic indicator in multi-range test meters. More sensitive meters with full scale current ratings of 25 or 30 microamperes can also be obtained. A further advantage is that this type of meter is relatively unaffected by external magnetic fields because the coil operates inside the strong magnetic field produced between the pole pieces of the internal permanent magnet.

The main disadvantage of a moving coil meter is that it requires a powerful permanent magnet and precise manufacturing techniques, which make it more expensive to produce than a moving iron type meter. Although moving coil meters are fairly robust it must be remembered that the coil is mounted in precise jewelled bearings so care must be taken not to expose the meter to unnecessary shocks, such as those caused by dropping it.

Unlike the moving iron meter, a moving coil type responds to the direction in which the current flows through the coil. The terminals of this type of meter are marked + and − to indicate the correct polarity for connecting the meter into a circuit. If the polarity is reversed, the meter needle is driven in the opposite direction and will try to go past the zero on the left hand end of the scale. This sort of treatment can result in the meter ending up with a bent pointer. Because the meter is sensitive to signal polarity it will not respond directly to alternating current signals.

Centre zero meters

Most moving coil meters have the zero position of the pointer at the left hand end of the scale. For some applications it is more convenient to have a meter which has its zero point at a point half way along the scale. This type of meter is known as a centre zero meter and is useful where the signal being measured can swing to either positive or negative values.

An example of an application for a centre zero meter is as the indicator in a DC bridge circuit. When the bridge is correctly balanced no current flows and the pointer will be at the centre of the scale. If the bridge is unbalanced in one direction a positive current flows through the meter and the pointer swings to the right of centre. If the bridge is unbalanced in the opposite direction a negative current (i.e. in the opposite direction) flows and the pointer swings to the left.

Centre zero operation is achieved by altering the setting of the tension springs so that the needle is moved to the mid scale position when there is no current flowing. This type of meter is specially designed for centre zero operation and a conventional left zero meter does not normally have enough zero adjustment range to allow the zero point to be shifted to mid-scale.

An alternative technique for achieving centre zero operation is to bias a conventional meter electrically by adding a bias current to the signal being measured so that with zero signal the meter reads mid scale. If this method is used a new scale may be added to show the centre zero operation.

Voltage measurement

For most electronic applications the parameter to be measured is the voltage between two points in the circuit. The moving coil meter itself is basically a current measuring device but it can be adapted for voltage measurement by making use of Ohm's law.

If we take a known resistor and connect it across the points where the voltage is to be measured the current flowing through the resistor is given by

$$I = \frac{V}{R} \text{ amps}$$

where V is the voltage across the circuit in volts and R is the resistance in ohms (Ω). The resistance is a constant so the current flowing in the circuit is directly proportional to the voltage applied.

If a moving coil meter is connected in series with the resistor as shown in Figure 1.3 it will measure the current flowing through

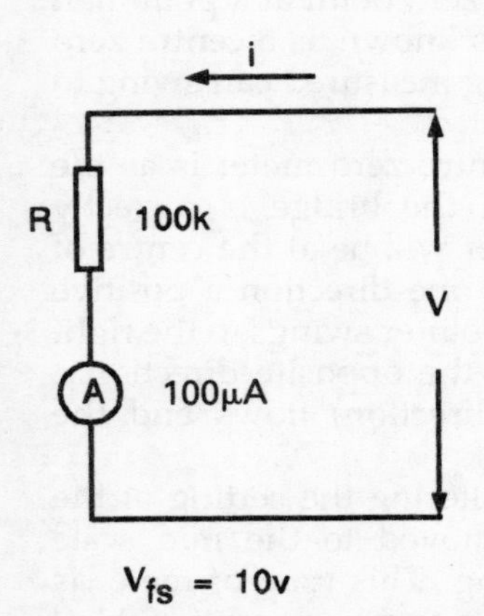

Fig 1.3 The basic voltmeter circuit.

the circuit. Because the voltage across the circuit is proportional to the current flowing through the resistance and the meter it is possible to calibrate the meter scale directly in volts. The meter has thus been converted into a voltmeter by adding a resistor in series with it. In this case it is assumed here that the meter coil resistance is zero and therefore has no effect on the current flowing in the circuit.

Multiplier resistors

To make up a voltmeter all we need to do is take a moving coil meter and connect a resistor in series with it. The test leads of the new instrument are now taken from the − side of the meter and from the resistor which is connected in series to the + terminal of the meter.

If we re-arrange the Ohm's law equation to get volts in terms of current and resistance we get

$$V = I.R$$

Now our moving coil meter can be calibrated in volts by simply muliplying its current reading by the value of the resistor in ohms. The series resistor used to convert a moving coil current meter into a voltmeter is usually referred to as a voltage multiplier resistor.

Suppose we have a basic meter with a full scale reading of 1mA and we wish to convert it into a voltmeter with a full scale reading of 1 volt. We need to choose a value of multiplier resistor so that when the voltage across the resistor and meter in series is 1 volt the current will be 1mA. The required resistance is given by the equation

$$R = \frac{V}{I} = \frac{1}{0.001} = 1000 \text{ ohms}$$

If we want to increase the full scale voltage reading to 10 volts the multiplier resistance needs to be increased by a factor of 10 so that it becomes 10k Ω. Now the full scale current of 1mA will flow when a voltage of 10 volts is applied across the voltmeter. By choosing a suitable value of series resistor the basic 1mA meter can be arranged to act as a voltmeter with any desired value of full scale voltage reading.

Ohms per volt ratings

So far we have looked at a meter based on a movement with a 1 mA full scale sensitivity. If a more sensitive meter is used so that the full scale current is only 100μA the value of series resistance required for a given full scale voltage reading will be higher than for a 1 mA movement. For a 100μA meter the series multiplier resistor required to give a 1V full scale reading would be 10000 Ω.

This brings us to an alternative way of defining the sensitivity of the basic meter movement by using the value of the series resistor required for a 1 volt full scale reading. Thus a voltmeter based on a 1mA full scale meter movement is often referred to as having a sensitivity of 1000 ohms per volt. If a 100μA meter is used the resultant volmeter would have a sensitivity of 10,000 ohms per volt and a 50μA meter gives a sensitivity of 20,000 ohms per volt.

Effect of coil resistance

When producing a voltmeter by using a microammeter and a series multiplier resistor the actual resistance of the meter coil will need to be taken into account if accurate scale calibration is to be achieved. This will become more important on low voltage ranges where the meter coil resistance can become significant compared with the value of the multiplier resistance.

If we consider a meter of 100μA full scale the coil resistance might be as high as 3000 Ω. If the meter were to be operated as a voltmeter with a 1 volt full scale range then the nominal value for the multiplier resistance would be 10,000 Ω. If we used the simple approach described earlier and wired a 10kΩ resistor in series with the meter the total resistance in the circuit would actually be 13kΩ due to the added resistance of the meter coil. In this case the real full scale reading for the meter would be 1.3 volts instead of the 1 volt that was originally intended. Thus ignoring the meter resistance would produce a scale error of the order 30% which would be unacceptable.

On higher voltage ranges the meter resistance becomes less important. For a 10 volt full scale range with a multiplier resistor of 100k Ω the error produced by ignoring the meter resistance is around 3%. When the same meter is set up for a 100 volt full scale reading the nominal multiplier resistance will be 1 MΩ. In this case the 3000 Ω meter resistance represents only 0.3% of the total series resistance and can be ignored since the error produced is negligible compared with the errors due to the accuracy of the meter itself.

To correct for the effect of the coil resistance, the multiplier resistor connected in series with the meter must be reduced in value by an amount equal to the meter coil resistance so that the total resistance in the circuit is correct for the voltage range required. Thus for our 100μA meter with its 3kΩ coil resistance the series resistor needed for a 1 volt full scale range will be

$$10000 - 3000 = 7000\ \Omega$$

The nearest convenient standard resistor value would be 6800 Ω which would produce a total multiplier resistance of 9800 Ω giving a scaling error of the order 2%. This is probably adequate for a meter intended for general use in servicing or for hobbyist applications. To achieve better results two resistors of say 6800 and 220 Ω could be connected in series to give a total resistance of 10020 Ω and a scaling error of the order 0.2%.

Electronic components suppliers often provide special precision multiplier resistors for use with their larger panel meters. These resistors have the meter coil resistance subtracted from the nominal multiplier value. The meters are scaled for both 0 – 3 and 0 – 10 ranges, and the appropriate resistors are available to make up meters with full scale voltage values of 3V, 10V, 30V, 100V and 300V. For other types of meter the series multiplier can be made up from one or two series connected standard value resistors which should be metal film types to give good temperature stability.

Measuring meter resistance

When a meter is obtained, the supplier's catalogue will usually give details of the coil resistance. Sometimes a meter may have the coil resistance marked on the dial face. If the coil resistance of the meter is not known it is not advisable to attempt to measure it by using an analogue multimeter set to its resistance range. The current produced in the external circuit by a multimeter, set to read resistance, can be quite high and may well damage the meter that you are trying to test. Even a digital multimeter can produce enough current to drive a sensitive meter beyond its full scale reading.

The resistance of the meter coil can be estimated reasonably accurately by using a simple test rig as shown in Figure 1.4. Here

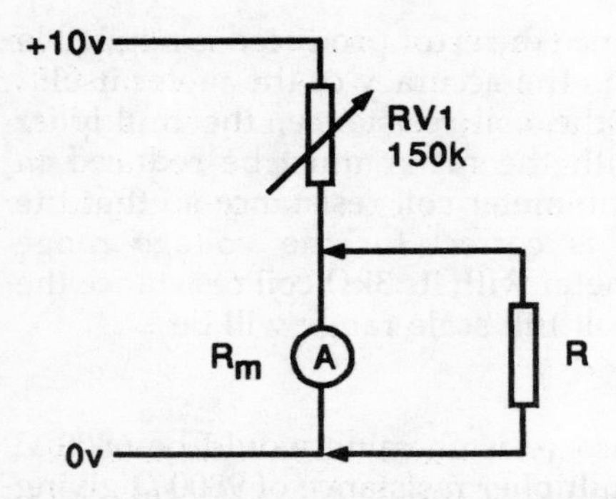

Fig 1.4 Circuit for measuring meter coil resistance.

the meter is fed via a variable resistor from a voltage source, such as a battery or power supply. The maximum value of the variable resistor should be chosen so that it will produce a current of about 60% of full scale on the meter. Before connecting the supply voltage the variable resistor must be set to its maximum value. With the supply connected the meter will produce a reading and the variable resistor should be adjusted to bring the meter needle up to full scale.

A series of resistors ranging from about 1kΩ down to perhaps 100 Ω are now connected, in turn, in parallel with the meter itself. When a resistor is connected across the meter it shunts some of the current away from the meter coil. If the new reading of the meter is above 50% of full scale the shunt resistor is larger than the coil resistance and a lower value should be tried. If the meter reading is 50% of full scale then the shunt resistor will be equal to the coil resistance. If an exact 50% reading is not obtained using standard resistor values it is possible to estimate the coil resistance by taking the resistance and scale reading for a value just above 50% and the resistance and scale reading for the value just below 50%. If the meter readings are roughly the same amount above and below 50% the average value of the two resistors can be used.

If a multimeter is available then a variable resistor of say 5kΩ could be switched in parallel with the meter under test and then adjusted until the meter reads half scale. The variable resistor can then be removed from the circuit and its value measured by using the multimeter.

This technique is not absolutely precise since the addition of the shunt resistor will also alter the voltage developed across the meter itself and this will slightly reduce the resultant meter reading. The average meter however is unlikely to have more than about 0.3V across it at full scale reading. If the supply voltage used is of the order 10V the effects of adding the shunt resistor are small and the value of meter coil resistance obtained by this method should be adequate for most purposes.

Simple multi-range voltmeter

The home constructor can produce a multi-range voltmeter as a fairly simple construction project. For the basic meter movement a 100μA moving coil meter is convenient and it should preferably be of the type with a scale length of the order 100mm. Apart from the meter, the components required are a number of resistors to provide the required voltage ranges and a case to contain the completed unit.

The simplest scheme for range selection is to have a separate 4mm socket for the positive connection of each voltage range with a common negative terminal as shown in Figure 1.5. Here a separate multiplier resistor is connected into circuit as each range is selected.

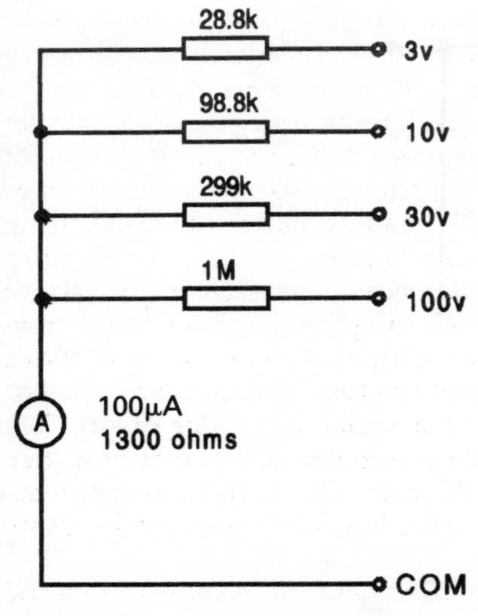

Fig 1.5 Simple multiple range voltmeter.

For a 100μA full scale meter movement the sensitivity rating is 10kΩ per volt and the series resistance required for each range is given by the full scale voltage required multiplied by 10k Ω. Thus for the 10V range the total series resistance should be 100k Ω. In the circuit shown the values have been corrected to take account of a coil resistance of 1300 Ω. The actual values needed will need to be adjusted to suit the meter being used for the project. If the resistors are of 1% tolerance the voltage readings produced will be in the order of 1% accuracy which is probably about the best that can be expected for a meter of this type.

In use this simple meter unit would have its negative test lead plugged into the common negative terminal and the positive test lead is then plugged into the appropriate socket for the voltage range desired. If the voltage to be measured is unknown it is advisable to start by using one of the higher voltage ranges and then to change down to the lower ranges until a sensible scale reading is obtained.

As an alternative to the plug and socket range selection scheme a more convenient arrangement is to use a rotary switch to select the voltage ranges. Now only two sockets are needed for the test leads and the circuit becomes as shown in Figure 1.6. Commercial

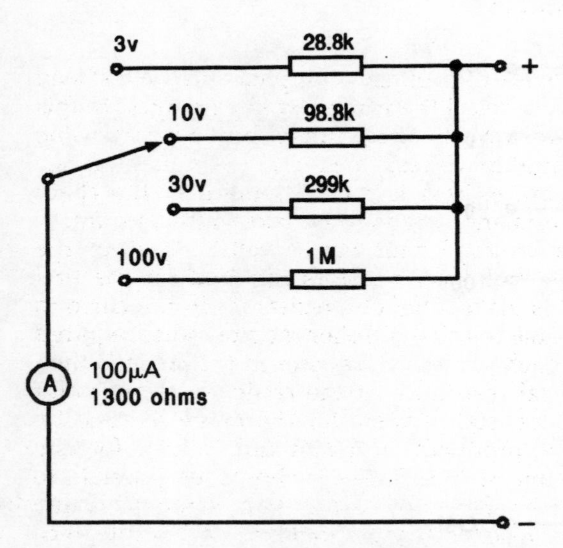

Fig 1.6 Voltmeter using switched multiplier resistors.

multimeters use specially designed switch assemblies to give low switch contact resistance and to provide good insulation between adjacent contacts when high voltage ranges are being selected. For voltages of a 1000V or more a separate socket for the test lead is desirable so that the high voltage is not applied to the range switch.

Another arrangement for the multiplier resistors, which may be more convenient when the meter resistance is to be taken into account, is shown in Figure 1.7. This scheme uses a chain of

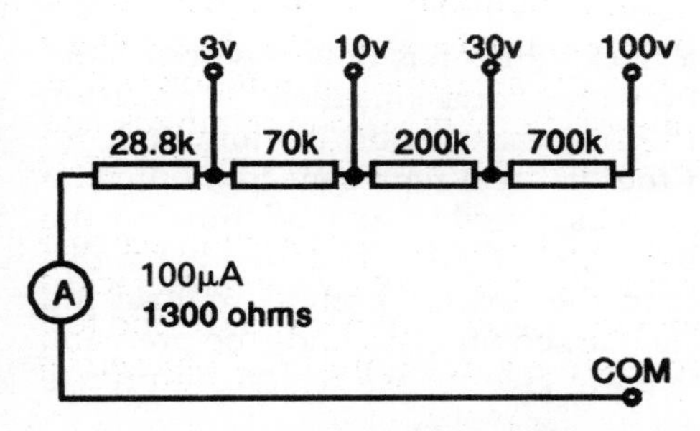

Fig 1.7 Voltmeter using a tapped multiplier chain.

resistors in series to provide the voltage multiplier and the test lead is tapped into the chain to select the various voltage ranges. In this case the resistance for the lowest voltage range should have a value equal to the nominal multiplier resistance for that voltage range minus the meter resistance. This is the only point in the chain where the meter coil resistance needs to be taken into account.

The second resistor in the chain has a value equal to the difference between the multiplier resistors required for the first two voltage ranges. This makes the total resistance in the chain to the second tap point equal to the multiplier resistor value required for the second range. Each successive resistor in the chain is then chosen to make the total resistance of the chain to its tap point equal to the multiplier resistance value for the range selected by that tap. This usually produces non-standard values for the resistors in the chain but in most cases it should be possible to produce the desired resistance by using two standard value resistors in series. The resistors used should be of 1% tolerance and preferably of the metal oxide type.

Accuracy and precision

It is important to understand the difference between the terms precision and accuracy as they are applied to measuring instruments. Although these words tend to be used to mean the same thing in general conversation they have different meanings when applied to a meter.

Precision is an indication of the degree of error that can occur in reading the scale of the meter. Accuracy is a measure of the error between the scale reading and the actual value of the quantity being measured.

If we take as an example a clock which has only the hour hand fitted and the dial is marked off with 12 divisions then the precision to which time can be read is to the nearest hour. If a minute hand is added then each mark on the dial will represent 5 minutes on the minutes scale and it becomes possible to read time to the nearest 5 minutes. If individual minute marks are added to the dial the time could then be read to the nearest minute. With a full sweep second hand and a scale marked off in seconds the precision of reading the clock is improved to give results which are precise to the nearest second.

The ability to read the time from the clock to the nearest second does not necessarily mean that the resultant time reading is accurate. If the clock has stopped the time will be correct at two moments during the day but all other readings of time would be meaningless. If the clock has been set so that it is five minutes fast we can read off a precise time but it will always be five minutes different from the real time so here we have an accuracy error of five minutes. If the clock is running fast so that it gains a minute per day then the accuracy of the readings of time will become progressively worse at a rate of one minute for every day after the clock was correctly set. Once again we can read a precise time from the clock dial but it is not the correct time.

If we apply the same principles to an analogue meter then the precision will be governed by the length of the scale and the number of calibration divisions on the scale. In simple terms if the meter has 100 divisions on its scale it should be possible to take readings to a precision of 1% of full scale. Thus if the full scale value is 100V we should be able to take readings to the nearest volt.

Apart from the scale markings there are other considerations which have to be taken into account when assessing the precision of a meter. The thickness of the pointer will affect the ability to read its position correctly. Another problem is that the pointer is usually mounted a small distance above the scale to ensure clearance and unless the meter is viewed perpendicularly to the plane of the scale there will be a parallax error in determining the pointer position. If the meter is viewed from the right side the needle position will appear to be lower on the scale whilst viewing from the left side will give an apparent reading which is higher than the true reading. This problem can be overcome by having a strip of mirror laid parallel to the scale itself. When the meter is read an image of the needle will be seen in the mirrored strip and this image should be aligned with the needle itself. When the two images are aligned the scale is being viewed at the correct angle and any parallax errors should be minimised.

When we consider a meter the term accuracy is a measure of the difference between the scale reading on the meter and the actual value of the voltage or current being measured. The accuracy is usually expressed as a percentage of the full scale reading of the meter. For a typical small panel meter the overall accuracy is likely to be of the order 2% to 2.5% of the full scale reading. For test meters, such as an Avometer, the accuracy is rather better and might typically be of the order 1% of full scale. Precision meters, specially designed for use in laboratory measurements, will usually have an accuracy which is better than 0.5% of full scale.

It is important to remember that the error in a meter reading may be either + or − relative to the true value. If we consider a simple panel meter with a full scale rating of 100 mA and an accuracy of 2% which shows a reading of 50 on its scale the true current flowing in the meter could have any value from 48 mA up to 52 mA.

In diagnostic work such a degree of error may be acceptable since many of the measurements taken are merely used as a general indication that a circuit is operating correctly. If the measurements are being taken as part of a laboratory experiment then the potential errors introduced by the accuracy rating of the meter accuracy and the precision of taking the readings must be allowed for when the results of the experiment are assessed. A fairly standard approach to this problem is to repeat the experiment several times and then to take an average result from the series of experiments.

Accuracy ratings for meters are normally specified at a nominal operating temperature of 20°C. If the meter is subjected to significantly higher or lower temperatures this can have an effect on the accuracy of the reading obtained.

Choice of range ratios

When making readings from a meter the best precision will be obtained when the meter reading is in the upper part of the scale. This factor becomes important when a choice is made for the series of ranges to be provided on a multi-range meter. This involves a trade-off between providing good precision and having the minimum number of separate ranges for voltage or current.

Suppose the meter were set up with a 10:1 ratio between ranges so that the full scale readings for the series of switched ranges might be 1V, 10V and 100V. If we want to measure a voltage of say 20V then this meter would have to be set to the 100V range and would give a reading which is 20% of full scale. If the meter has 50 scale divisions the voltage can be read to the nearest 2V which represents a potential error of 10% relative to the 20V value being measured.

By adding two intermediate ranges of 5V and 50V full scale it becomes possible to measure the 20V signal using the 50V range. This gives a meter reading which is at 40% of full scale and allows the voltage to be read off to the nearest 1V which reduces the potential error on the 20V value to 5%.

In commercial multi-range meters the ranges may be arranged in the sequence 1, 2, 5, 10, 20, 50, 100 and so on. This scheme allows meter deflections of at least 40% of full scale to be obtained for any voltage by selecting an appropriate range on the meter. For most voltages when the optimum range is selected the reading will be in the upper half of the scale.

An alternative scheme that is often used in multi range meters follows the sequence 1, 3, 10, 30, 100 and so on. This requires a smaller number of different ranges and ensures a reading of between 30% and 100% when the optimum range is selected. Another scheme which is sometimes used on multimeters has the ranges following a 3, 6, 12 sequence.

Meter protection

When a moving coil meter is set up as a multi-range voltmeter it is unfortunately very easy to connect the meter test leads to a high voltage circuit whilst the meter has a low voltage range selected. The result is that the meter is subjected to much more than its full scale current and could be badly damaged or even destroyed.

It is always good policy to start a new measurement with the meter set on its highest voltage range and then to select lower ranges as required until a sensible reading is obtained.

Moving coil meters are generally designed so that they can withstand overloads up to twice their full scale rating without being damaged. In fact most meters can stand overloads up to ten times full scale for a brief period. Precision meters usually have small buffer springs at each end of the pointer movement range and these absorb the impact of the pointer to reduce damage when the meter is overloaded. Extreme overloads can nevertheless cause the pointer needle to become bent and in the worst case the coil itself could be burned out thus destroying the meter.

To avoid serious damage to the meter all commercial multi-range meters are fitted with protective circuits. The usual arrangement is to have a pair of silicon diodes wired across the meter itself as shown in Figure 1.8. Under normal conditions the voltage drop

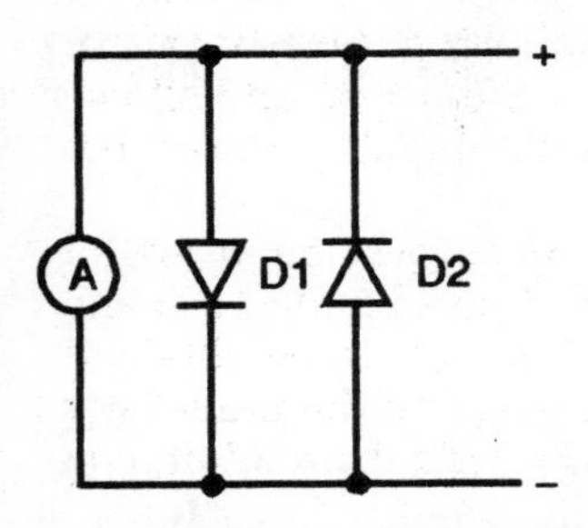

Fig 1.8 Meter overload protection using diodes.

across the meter coil at full scale current is about 200 – 300 mV. Silicon diode D1 will start to conduct when the voltage across the coil reaches about 500 mV and as the voltage rises beyond this level most of the current will be diverted through the diode thus protecting the meter coil. The second diode protects the meter when a reverse voltage is applied to the meter.

Multimeters such as the Avometer Model 8 use a different protection technique. This makes use of a sensitive cut-out relay which breaks the circuit to the meter when excessive current flows. On this type of meter a small button on the meter case pops up when the cut-out circuit is tripped by an overload. After the cause of the overload has been rectified the button can be pushed in again to reset the cut-out and reconnect the meter circuit.

Current measurement

The measurement of direct current is, in theory, quite straightforward since the meter movement itself responds to the current flowing through its coil and the needle deflection is directly proportional to the amount of current flowing. For relatively small values of current, up to that which produces full scale deflection of the needle, the meter can be connected in series with the circuit in which the current is to be measured. This approach may be used for panel meters where the meter sensitivity is chosen to cover the maximum amount of current that is expected in the circuit.

For a general purpose test meter which is to be used for both voltage and current measurements the basic meter movement is usually chosen to have a high sensitivity so that its full scale current rating will be small. In this case the current to be measured is likely to be much greater than that for which the basic meter is designed. In order to prevent damage to the meter a different approach must be adopted to provide the ability to measure higher current levels.

Current shunt resistors

Suppose the basic meter we are using has a full scale current range of 1mA and we wish to measure current values up to 100mA. Connecting the meter in the normal way would cause the full circuit current to flow through the meter coil and would probably result in its destruction.

To avoid this situation we must use a technique where most of the current is diverted around the meter so that no more than 1mA is actually allowed to flow through the meter coil. This can be achieved by connecting a shunt resistor across the meter coil as shown in Figure 1.9.

Suppose we arrange that the resistor connected in parallel with the meter has the same value of resistance as the meter coil itself. For a given voltage across the circuit the current flowing in the shunt resistor will be the same as that flowing through the meter coil. The total current flowing through the circuit is therefore twice that flowing through the meter. When the meter reads full scale the total current flowing in the circuit is 2mA so we have effectively doubled the full scale rating of the meter. The meter scale could now be re-calibrated to show 2mA as the full scale reading.

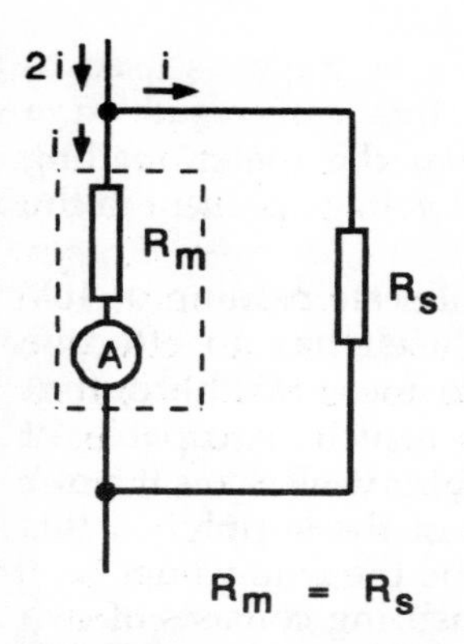

Fig 1.9 Ammeter circuit using a shunt resistor.

To make the 1mA meter operate as a 10mA full scale meter we need to have 1mA flowing through the meter, to provide its full scale reading, and the remaining 9mA must be diverted through the shunt resistor. If the shunt resistor is to take nine times the current flowing in the meter coil its resistance must be equal to 1/9 of the resistance of the meter coil.

The same basic principles can be applied to calculate shunt resistors for any desired value of full scale current. So for the 100mA full scale current meter we would require a shunt resistor with a value equal to 1/99 of the meter coil resistance, and, for 1A, the shunt resistor becomes 1/999 of the coil resistance.

The formula for calculating the shunt resistor is

$$R_s = \frac{R_m . I_m}{I_{fs} - I_m} \text{ ohms}$$

where R_m is the meter coil resistance in Ω, R_s is the shunt resistance, I_m is the meter current rating and I_{fs} is the desired full scale current range for the meter plus its shunt. The current ratings I_m and I_{fs} must be in the same units which may be either milliamperes or amperes.

Meter loading

An important problem with a simple voltmeter, based on a moving coil type movement, is that the meter circuit must draw some current in order to move the pointer. When the meter is used to measure voltage in an electronic circuit this can result in a disturbance to the circuit conditions so that the meter reading obtained is not a true indication of the actual voltage present in the circuit.

Suppose we are using a meter with a full scale reading of 10V based on a 1 mA movement. The meter itself has an effective resistance of 10kΩ. If the meter is used to measure the output voltage of a 5V stabilised power supply which is capable of delivering up to 1A of output current the meter itself takes 0.5 mA which has no effect on the output voltage of the supply. In this case the voltage reading on the meter will be the true value.

Now suppose that the circuit we are measuring consists of two 5kΩ resistors, R1 and R2 in series across a 10V supply as shown in Figure 1.10. The meter is then connected across R2 in order to

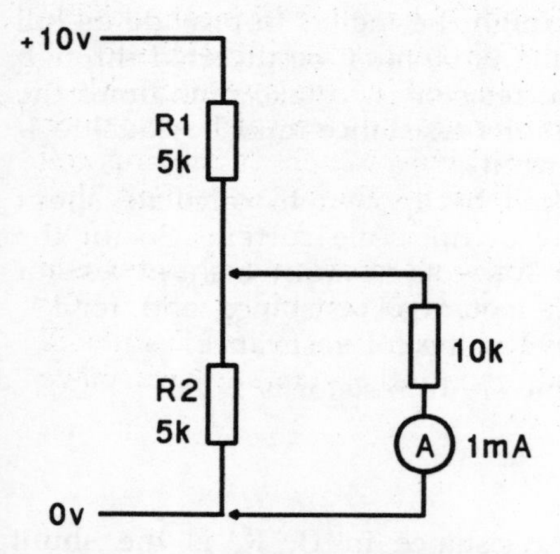

Fig 1.10 Circuit demonstrating voltmeter loading effect.

measure the voltage at the junction point relative to the bottom of the chain. When the meter is not connected the voltage at the junction point will theoretically be 5V since the two resistors are equal in value and carry the same current.

When the meter is connected to the circuit it effectively places a 10kΩ resistor in parallel with R2 and some of the current is diverted from R2 into the meter circuit. The result of adding the meter in parallel with R2 is that the effective value of the lower part of the resitor chain is now reduced to

$$R = \frac{5000 \times 10000}{15000} = 3333 \text{ ohms}$$

This will cause the voltage at the junction point to be reduced when the meter is connected and the meter will give a reading of

$$V = \frac{10.R}{R1+R} = \frac{10 \times 3333}{8333} \; 3.99 \text{ volts}$$

This is about 1V lower than the voltage that would have existed with no meter connected. The effect of adding the meter is to produce a 20% error in reading the voltage at the junction of R1 and R2.

The higher the resistance of the meter the smaller will be its effect on the voltage reading in the circuit. When measurements of voltages within a working circuit are to be made it is therefore desirable that the voltmeter used should have the highest possible resistance so that its loading effect on the circuit is minimised.

There is a limit to what can be achieved by this approach. More sensitive meters are expensive to manufacture and also tend to be rather more fragile than a low sensitivity meter. A moving coil meter of 50μA full scale current is about the limit of sensitivity that is practical for general use.

A voltmeter based on a 50μA full scale moving coil meter is a good choice since this gives 20kΩ per volt sensitivity. On a 10V range such a meter will have a loading resistance of 200kΩ and will give reasonable accuracy when used on circuits which have impedances of less than about 20kΩ.

The electronic voltmeter

An alternative approach to dealing with the problem of circuit loading caused by connecting a voltmeter into a circuit is to make use of an electronic amplifier to provide the current which drives the meter. By using a suitable circuit the input impedance of the electronic amplifier can be made extremely high so that the input current drawn by the amplifier is negligible. When this type of meter is used it is possible to make measurements on circuits which have quite high impedance without seriously disturbing the voltages present across the part of the circuit being checked. This type of meter is called an electronic voltmeter.

Early types of electronic voltmeter used a valve to provide the amplification and the instrument was referred to as a valve voltmeter (VVM) or vacuum tube voltmeter (VTVM).

A typical circuit arrangement for a simple valve voltmeter is shown in Figure 1.11. A pair of matched triode valves are used and

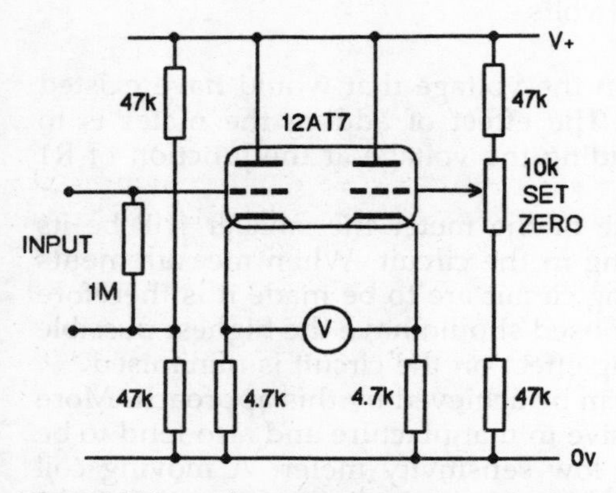

Fig 1.11 Valve voltmeter circuit using a dual triode.

each valve operates as a cathode follower. The grid of the first triode V1 is fed by the input signal and its cathode voltage changes in sympathy with the input signal to give an amplification of roughly unity. The second triode is fed from a fixed bias voltage and its cathode provides a fixed reference against which the input signal is measured. The negative input test lead is also tied to the bias voltage for V2.

When an input voltage is applied to the grid of V1 the cathode voltage of V1 rises to produce a voltage difference between the two valve cathodes which is approximately equal to the input voltage. The moving coil meter and its associated multiplier resistor is connected between the cathodes and will provide a voltage reading corresponding to the input voltage. The current required to drive the meter is now derived from the anode to cathode current flowing through the two valves.

The effective input impedance of a cathode follower circuit is very high compared with the cathode load resistance since the circuit forms a negative feedback amplifier with virtually 100%

feedback. In theory the input resistance of such a voltmeter would be infinite but in practical circuits it is likely to be of the order 1 – 2 MΩ.

The bias level applied to the grid of V2 is made variable so that the meter reading can be set to zero when the input terminals are shorted together. This balances out differences in the characteristics of the two triodes. In a practical circuit the valve used would be a dual triode, such as the 12AT7, where the two triodes are built into the same envelope and the characteristics of the triodes are closely matched.

The valve voltmeter has the disadvantage that it requires a power supply which tends to limit its use as a portable instrument. The valves can also generate quite a lot of heat so the instrument needs to be switched for a time to allow the effects of temperature changes inside the case to settle down to a stable level before any serious measurements are taken.

The modern equivalent of a valve voltmeter might use junction field effect transistors in place of the triode valves as shown in Figure 1.12. This circuit has similar characteristics to a valve

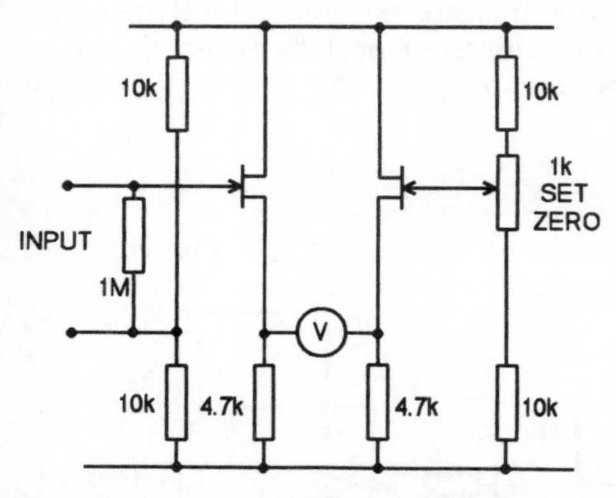

Fig 1.12 Electronic voltmeter using junction FETs.

voltmeter except that the supply voltage is lower and could be provided by a battery. The transistor version consumes very little power in comparison with the valve circuit so that problems with changes in temperature are reduced. The low power consumption means that an FET voltmeter circuit can readily be built as a portable instrument. The voltage range that can be measured by the basic instrument is limited to perhaps 5 – 10V because of the voltage ratings of the transistors. To provide higher voltage ranges a potential divider can be added in front of the basic voltmeter to reduce the input voltage to an acceptable level. The addition of the potential divider will reduce the effective input impedance of the instrument but this can still be of the order several megohms which is much higher than the impedance of a conventional moving coil voltmeter.

Integrated circuit voltmeter

Although electronic voltmeters using discrete transistors can quite easily be produced a more convenient technique is to use a pair of integrated circuit operational amplifiers to drive the meter. A typical circuit arrangement for such an instrument is shown in Figure 1.13.

Both amplifiers are operated as voltage followers where the output is fed back directly to the inverting input terminal of the amplifier. Because this connection provides 100% negative feedback and the internal amplifier has an extremely high gain the circuit produces an overall amplification of 1 between the non-inverting input terminal and the output. The result is that the

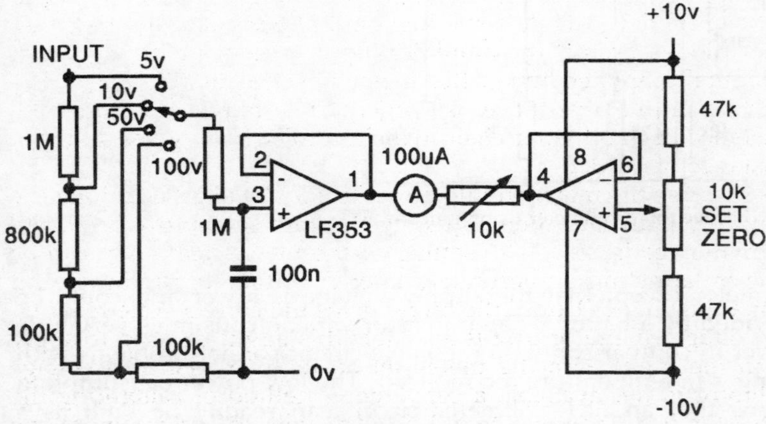

Fig 1.13 Electronic voltmeter using LF353 amplifier.

output voltage is the same as the voltage applied to the non-inverting input of the amplifier. Another advantage of the negative feedback is that the amplifier has a very high input impedance. The signal to be measured is applied to the non-inverting input of the first amplifier and the input of the second amplifier is fed from a variable bias voltage. The meter and its multiplier resistor is then connected between the outputs of the two amplifiers and these provide the current needed to drive the meter.

In the circuit shown, the LF353 device used contains two operational amplifiers and these use junction field effect transistors (JFETs) in their input circuits to give a very high input impedance. A simple low pass filter is included between the meter input and the amplifier input to remove any noise or mains ripple that might be picked up on the test leads feeding the high impedance input. The bias applied to the second amplifier is made adjustable and is preset to make the meter read zero when there is no input applied to the circuit. For a mains powered instrument the power supply lines could be +10V and −10V and would normally be stabilised to give constant voltage. A battery powered instrument might use two 9V batteries connected in series as the power supply.

The multiplier resistor used in series with the meter can be adjusted to give the desired full scale voltage. For higher voltage scales an input voltage divider is used to scale down the input voltage to give a maximum of 5V input to the amplifier at full scale. In the circuit shown the basic range is set for 5V and the divider provides 10, 50 and 100V ranges. By altering the component values in the divider chain, other sequences of voltage ranges could be provided. The 1MΩ series resistor and shunt capacitor at the input to the amplifier provide some low pass filtering and the two 5.6V zener diodes protect the amplifier input in case a high voltage is accidentally applied when the meter is set at its 5V range. The effective input resistance is largely determined by the total resistance of the divider chain which in this circuit gives an input resistance of 10 MΩ.

Before use the meter should be set to zero by short circuiting the input terminals and then adjusting the zero set control RV2 until the meter reads zero. The setting of the meter calibration can be done by applying an accurately known voltage to the input and then adjusting RV1 until the meter gives the correct scale reading. If a voltage calibrator unit is available this might be used to apply 5V to the input with the meter set on its 5V range. If a voltage calibrator is not available a convenient method of calibrating the voltmeter is to connect two 1.5V dry cells in series across the voltmeter input. If the cells are new the voltage produced across the two cells should be 3.1V. With the 5V range selected RV1 is then adjusted until the meter reads 3.1V. The meter is then calibrated and ready for use. The variable resistor RV1 could be an internal preset control since it should not require frequent adjustment. The zero set control RV2 should be brought out to the front panel since this may need to be adjusted from time to time to compensate for temperature changes and variations in the instrument supply voltages if the unit is battery powered.

Measurements in AC circuits

In an alternating current circuit there are various ways in which the voltage or current can be specified. These levels are shown in Figure 1.14. We could measure the maximum change in voltage or

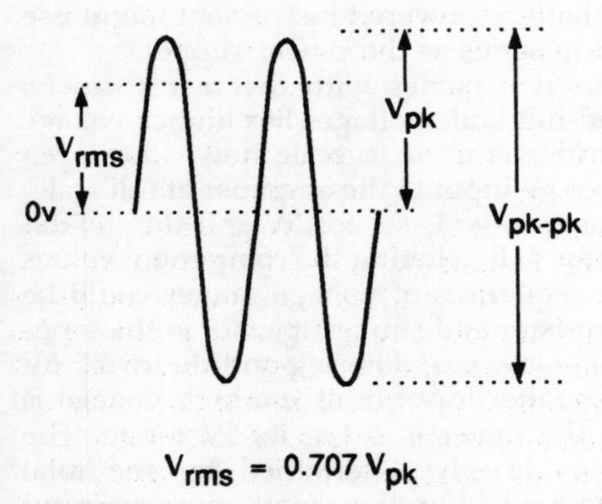

Fig 1.14 Voltage levels in an AC signal.

current relative to the zero level. This would be the level when the signal is at either its maximum positive or maximum negative negative point. This level is called the peak voltage V_{pk} or peak current I_{pk}.

An alternative measurement which is frequently used is the peak to peak voltage (V_{p-p}) or current (I_{p-p}). This is the total voltage swing from negative peak level to positive peak level of the waveform. This measurement is more convenient when voltage levels are being measured on an oscilloscope.

In practical applications current and voltage in an AC circuit are usually measured as a root mean square (RMS) value. This value gives the correct result when the formula $W = I^2.R$ or $W = V^2/R$ is used to calculate the power that would be dissipated in a resistor connected across the circuit. For a sine wave signal the RMS value is 0.707 times the peak voltage or current. It should be noted that for other waveforms the relationship between the peak and RMS levels will be different and may affect the calibration of a meter used to measure AC voltages or currents.

Measuring AC voltage

The conventional voltmeter based on a moving coil movement responds to a direct current (DC) signal. There are many applications where we may need to be able to measure an alternating current (AC) voltage.

If we take the a 50Hz signal as an example, the instantaneous value of the voltage follows a sine waveform in which each cycle has a period of 20ms. For one half of the cycle the voltage is positive and for the remaining half cycle it will be negative. The two halves of the sine wave are of the same amplitude and over the complete cycle the average value of the signal is zero.

If the meter movement were able to respond very rapidly so that the needle could follow the instantaneous variations in voltage then the needle would swing back and forth at a rate of 50 times a second, but its mean position would be at the zero point on the scale. In a typical meter movement the response is much slower so that the needle moves by only a very small amount on each half cycle and will appear to remain, more or less, at its normal zero position. In fact the meter tends to respond to the average level of the signal over a period of time and in the case of a sine wave this average level is zero.

To produce a sensible meter reading of the amplitude of an AC signal we need to convert the signal into an equivalent DC level which can then be applied to the meter to produce a reading which is proportional to the amplitude of the AC signal.

Diode rectifier meter

An alternating signal can be converted into a uni-directional signal by passing it through a rectifier diode as shown in Figure 1.15.

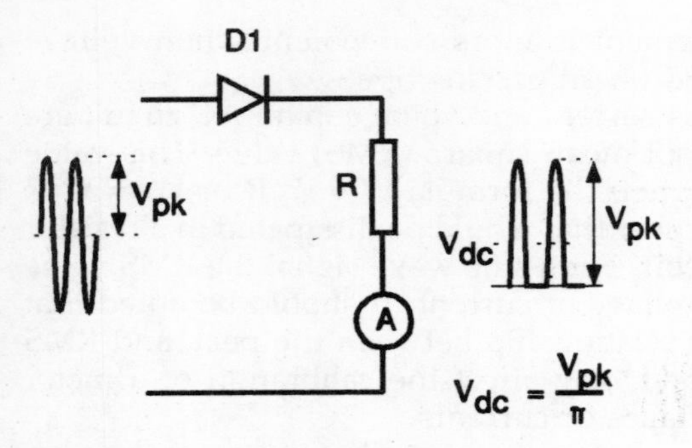

Fig 1.15 AC voltmeter using a diode rectifier.

Here the diode conducts during the positive half cycle of the signal but presents a high impedance to the negative half cycle. The signal applied to the meter now consists of a series of positive half cycles of the signal waveform with the signal at zero during the period of the negative half cycles.

The meter movement is unable to respond to the rapidly changing signal and gives a reading which is proportional to the average value of the signal over a period of time. For the series of half sine wave pulses the average value is

$$V = \frac{V_{pk}}{3.142}$$

which is approximately one third of the peak amplitude of the pulses. The meter can readily be scaled to read directly in RMS values by adjusting the value of the series multiplier resistor.

Bridge rectifier meter

Although a simple half wave diode rectifier can be used to make a meter read an AC signal the more commonly used scheme employs a bridge rectifier system as shown in Figure 1.16. The diodes of the bridge effectively switch the connections of the meter so that current always flows through the meter in the same direction on both half cycles of the waveform. During the positive half cycle current flows through diodes D1 and D4 to provide current through the meter. When the negative half cycle occurs these diodes switch off and current passes through diodes D2 and

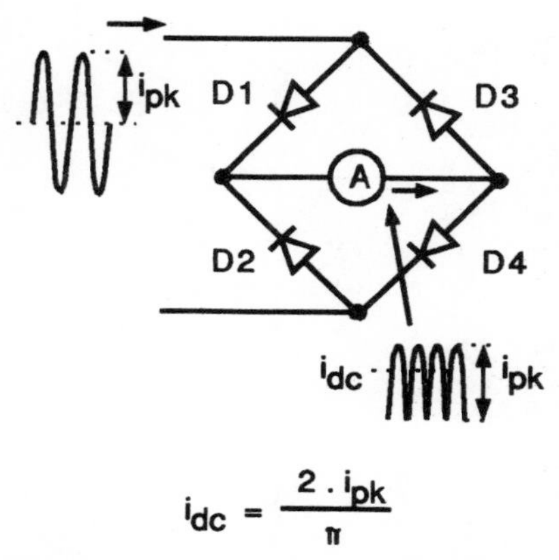

$$i_{dc} = \frac{2 \cdot i_{pk}}{\pi}$$

Fig 1.16 Meter using a bridge rectifier.

D3 to produce the same direction of current flow through the meter. Because the current flows in the same direction through the meter on both half cycles the average current flowing through the meter is twice the level produced by a simple diode rectifier circuit and the meter becomes twice as sensitive.

A typical meter rectifier is normally of the copper oxide type since this has a lower forward resistance and a lower voltage drop than either silicon or germanium type junction rectifiers. A meter using this type of rectifier can generally be used over a wide range of frequencies from about 40 Hz up to 10 kHz. If the meter is to operate at higher frequencies then silicon junction diodes would be used since they have lower capacitance than a copper oxide type. The problem with using silicon diodes is that the low end of meter scale will become more non-linear.

When the meter is used as an AC voltmeter the multiplier resistor is connected in series with the input side of the bridge rectifier as shown in Figure 1.17. This avoids the application of high voltages to the rectifier. A meter using the full wave rectifier circuit produces a DC voltage reading which is approximately 0.6 times the peak voltage of the signal applied. To make the scale readings correspond to RMS values the multiplier resistor value needs to be adjusted to produce the higher reading corresponding to the RMS value.

One problem with the rectifier type meter is that at low signal levels the voltage drop across the rectifier reduces the current

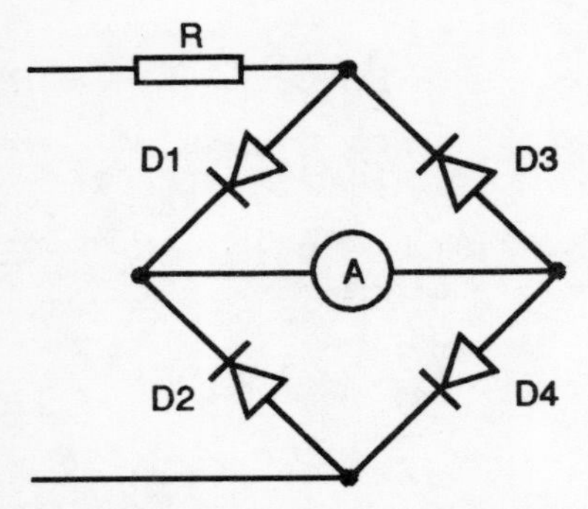

Fig 1.17 AC voltmeter using a bridge rectifier.

flowing in the meter and the calibration will be non-linear at this end of the scale. This effect is worst on the low voltage ranges and becomes less significant when the meter is used to read high voltages such as those on the supply mains.

The relationship between the RMS value of a signal and the peak, or peak to peak, value depends upon the waveform of the signal being measured. For a sine wave the RMS voltage is $0.707V_{pk}$ and this is the value normally used when calibrating the scale of a rectifier type meter. If the waveform is a square wave the RMS voltage becomes equal to V_{pk} and a normal AC meter driven by this waveform would give an incorrect reading. Other waveforms will also tend to give incorrect readings when applied to a rectifier type meter.

Peak reading meter

A meter can be connected so that it will read approximately the peak value of the voltage applied. In this case the signal is applied to a half wave diode rectifier which has a large capacitor connected across its output as shown in Figure 1.18.

During the period when the diode conducts the capacitor charges up to the peak voltage of the signal. If the load across the capacitor is of very high impedance and the reverse resistance of the diode is also high, the capacitor is unable to discharge during the period where the diode is turned off so it remains charged to the peak value. A voltmeter circuit connected across the capacitor can now be used to measure the peak voltage of the AC input

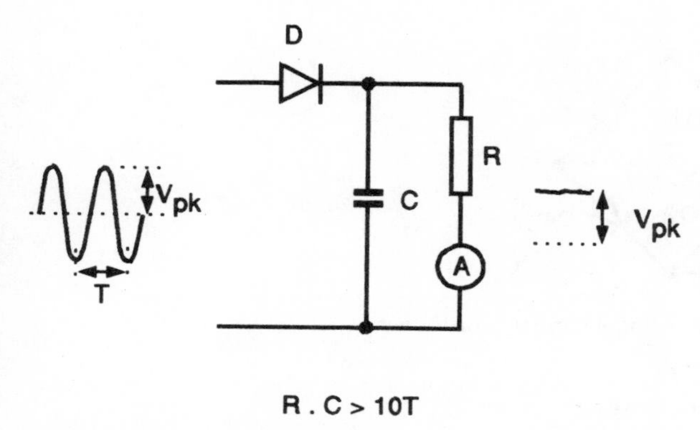

Fig 1.18 Peak voltmeter circuit.

signal. In practice the capacitor will discharge a little through the meter circuit so there will be a small level of variation or ripple on the capacitor voltage. If the product of capacitance C and meter circuit resistance R is made equal to about 10 times the time period of the AC signal the capacitor voltage will be virtually equal to the peak voltage of the signal. For this type of measurement an electronic voltmeter which has high input impedance will provide the most accurate results.

The capacitor used for low frequency signals will probably need to be an electrolytic type and for this purpose a tantalum type capacitor should be used to obtain the minimum leakage current. In radio frequency circuits a ceramic plate capacitor is generally used.

Current transformers

When a meter is used to measure current in an AC circuit and it is desired to extend the current range the technique of using shunt resistors across the meter is not a practical proposition because the meter rectifier would then have to carry the full circuit current.

For high current AC ranges a current transformer is usually employed as shown in Figure 1.19. This transformer has a low number of turns on its primary winding which is connected in series with the circuit carrying the current that is to be measured. The secondary winding of the transformer has more turns and

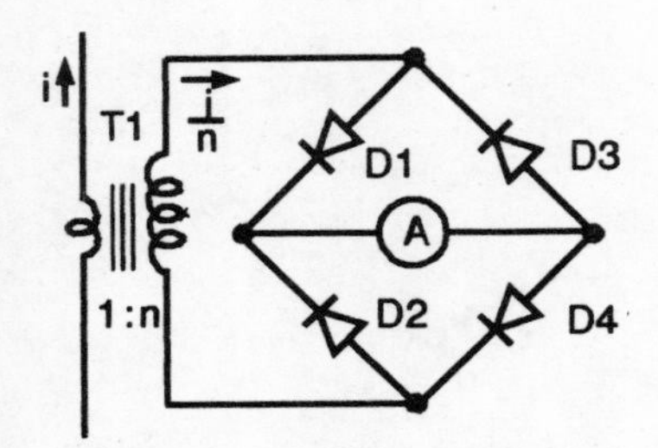

Fig 1.19 AC ammeter using a current transformer.

provides a step up ratio for voltage but a step down ratio for current. Thus if the turns ratio were 1:10 the current in the secondary winding would be 1/10 of that in the primary and this smaller current is used to drive the meter.

Clip around meters

When dealing with power circuits it may be an advantage to be able to measure the current flowing in a conductor without actually breaking into the circuit. When an AC circuit is to be tested this can be achieved by using a clip-around current meter.

The meter unit itself has a probe in the form of a iron or ferrite core which can be opened to allow it to be clipped over the current carrying conductor. When the clip is closed it forms the magnetic core of a transformer. The current carrying conductor itself now passes through the core and acts as a single turn primary winding. Another multi-turn winding on the core acts as the secondary of the transformer.

In this type of meter the clip on probe section acts as a current transformer and a rectifier type ammeter is connected to the secondary winding on the core. The current in the secondary winding is proportional to that flowing in the conductor by the ratio 1/n where n is the number of turns on the secondary winding. The meter is calibrated to show the current in the conductor which passes through the probe.

In a practical meter the use of the clip on core tends to reduce the meter accuracy slightly but typically readings of around 5% accuracy can be achieved. Meters of this type are normally designed for high power circuits and will usually have current ranges from 10 to 100A. By using a modified scheme in which the transformer has extra windings and is operated as a magnetic amplifier it is possible to measure current in DC circuits as well and some commercial meters of this type provide this facility.

Thermocouple meters

Although most of the meters which the reader is likely to encounter use the moving coil type movement, there are one or two other types of meter which may be used in certain special applications.

The thermocouple meter uses an indirect method of measuring current. The actual current being measured is fed through a heater element which is attached to a thermocouple temperature sensor element. The thermocouple generates a voltage which is proportional to its temperature and this voltage can be directly related to the power being dissipated in the heater element. The voltage output of the thermocouple can also be calibrated in terms of the current flowing through the heater. The measurement of thermocouple voltage is made by a conventional moving coil meter. In a typical unit the heater and thermocouple are built into the meter case and the scale is calibrated directly in terms of voltage or current in the heater circuit.

The advantage of a thermocouple meter is that it can read either AC or DC. This type of meter is also useful for measurements of radio frequency power since it is not frequency sensitive. Its main disadvantage is that the heater needs to absorb some power from the circuit under test in order to produce a meter reading.

Electrostatic voltmeters

This type of meter is used for high voltage measurements, above 1000V, and makes use of the force produced when a large electrostatic field is applied between two plates. One plate is fixed and the other is free to move in much the same way as the iron elements in a moving iron meter. The pointer is linked to the moving plate.

Electrostatic meters take virtually no current, which makes them useful for measuring high voltages such as those in the EHT supply for the picture tube in a television receiver. Their application tends to be rather specialised and the average engineer will rarely need to use this type of meter.

Power measurement

In a DC circuit the power can be measured by simply measuring the voltage across the load and the current flowing through the load and then multiplying the results together to give the power in watts. This requires the use of two meters but will work for any value of load resistance.

In an AC circuit if the load is resistive the power can again be measured by using two meters but voltage drop in the AC ammeter may be important so this type of measurement is mainly suitable for measuring power at the mains voltage level. If the load impedance has a significant reactive component, the power reading produced by the two meter method is likely to be wildly inaccurate unless the phase angle between voltage and current is taken into account. This phase angle could be determined by using an oscilloscope to display the current and voltage waveforms in the circuit.

True wattmeters

The measurement of the power consumption of a circuit can be carried out by using a special type of meter called a wattmeter. In this type of meter there are two active coils one being fixed and the other moving. Because the fixed coil determines the strength of the field in which the moving coil operates the needle movement is proportional to the product of the currents in the two coils. If the fixed coil is connected across the circuit being measured so that it responds to the voltage level and the moving coil is in series so that it responds to the current flowing, the movement of the needle is proportional to the product of current and voltage. This type of meter is calibrated to read power directly in watts.

Wattmeters are relatively expensive instruments and are mainly used in power engineering since they are usually inaccurate at the power levels likely to be met in electronic circuits. Thus a true wattmeter might be used to determine the power consumption of a piece of mains driven equipment but there are other methods of achieving the same results so the average electronics engineer is unlikely to need to use a true wattmeter. The main advantage of a dual coil wattmeter is that it will give correct power readings in an AC circuit where the load has a large reactive component.

Pseudo wattmeters

So far we have seen that it is possible to measure power in a resistive load by using separate meters to measure voltage and current. It is however possible to measure power in a resistive load using only one meter provided the resistance value of the load is known.

From Ohms law we know that V = I.R and we also know that the power is given by the product of V and I. Thus we get

$$\text{Power } W = V.I$$

and by substituting I.R for V we can get an equation for power in terms of current I and resistance R which is

$$W = I^2.R$$

Alternatively we can get a value for power in terms of voltage V and resistance R which is

$$W = \frac{V^2}{R}$$

If we now connect a voltmeter across the load R and measure the voltage then the power can readily be calculated. In fact for any given value of load resistance we could calibrate the meter directly in terms of power in watts. The resultant scale will be non-linear since the power is proportional to the square of the voltage measured. Similarly an ammeter or milliameter connected in series with the load will allow the power in the circuit to be calculated from the current reading. For a given value of load resistance the scale of the meter could be calibrated directly in watts or milliwatts as desired.

This approach is generally adopted for audio output power meters. The meter used is a rectifier instrument so that it can measure AC voltage or current and the scale is calibrated directly in watts. An alternative calibration might be in decibels relative to 1 watt. This type of meter is sometimes referred to as a pseudo wattmeter.

One problem with a simple wattmeter of this type is that the calibration is only correct for a particular value of load resistance. Commercial power output meters are usually designed so that they can be used with a number of different values of load resistance. This is achieved by using a matching transformer between the load and the meter so that the meter effectively sees the same load resistance at all times and therefore its calibration remains constant. The output meter may also include an amplifier circuit to provide a suitable signal for operating the indicating meter.

When power is being measured in an AC circuit where the load is reactive the pseudo wattmeter will give inaccurate readings since it will be reading the VA value and not the true power in watts. For most applications where power output is to be measured the load will be resistive. In the case of an audio amplifier output the loudspeaker could be replaced by an equivalent resistance when the output power is to be measured.

Potentiometric measurements

When a meter is used to measure a voltage level we have seen that the meter circuit needs to draw current from the circuit being measured in order to produce the needle movement. By using an electronic amplifier we have seen that it is possible to provide a very high input impedance which means that the current drawn can be negligible. Another method of achieving zero current measurements is the potentiometric method in which a conventional meter is used in conjuntion with a calibrated potentiometer.

The basic arrangement for this type of measurement is shown in Figure 1.20. Here a precision potentiometer RV1 is driven by current from a battery via an adjustable resistor RV2. Suppose the potentiometer has a total resistance of 10kΩ and the current flowing in it is adjusted to be exactly 1mA. The total voltage across

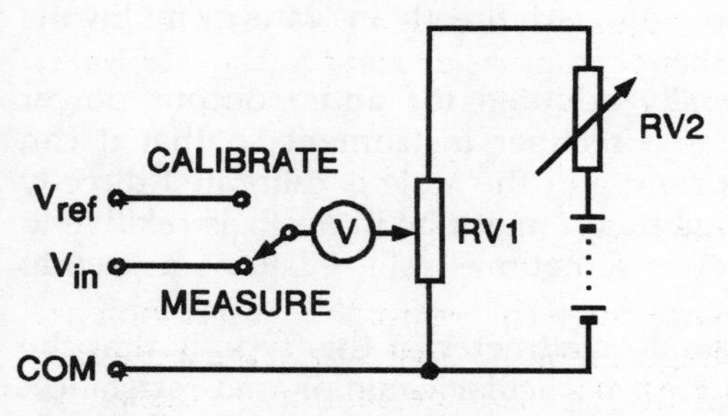

Fig 1.20 Voltage measurement using potentiometric method.

the potentiometer is now exactly 10V. If the resistance of the part of the potentiometer from the zero end to the wiper is R1 and the total resistance of the potentiometer is R2 then the voltage at the wiper relative to the zero end is given by the equation

$$V = \frac{10.R1}{R2}$$

The voltage at the wiper is therefore directly proportional to the position of the wiper and the potentiometer scale could be directly calibrated in terms of volts. This voltage is applied to one side of a centre zero meter and an unknown voltage is applied to the other side of the meter. If the potentiometer is now adjusted so that no current flows through the meter the voltage from the potentiometer must be exactly equal to the unknown voltage. Now the value of the unknown voltage can be read off from the potentiometer scale.

To set up the calibration of the potentiometer the switch is set to the calibrate position and a known voltage is applied to the calibrate input terminal. This voltage might be derived from a calibrated voltage source or perhaps a precision Zener diode. The potentiometer is then set to a point on its scale corresponding to that voltage. The variable resistor RV2 is used to adjust the current in the potentiometer until the meter reading is zero and the voltage scale on the potentiometer will then be correctly calibrated.

To make a measurement the switch is set to the measure position and the voltage to be measured is applied to the test terminal. The potentiometer setting is adjusted to produce a zero current reading on the meter and the unknown voltage can be read directly from the potentiometer scale. For a simple potentiometric measuring scheme the potentiometer itself could be one of the ten turn precision types which can be fitted with a precision dial to allow accurate readings of the wiper position to be made.

Since the meter is always adjusted to give a zero current when a measurement is taken, this type of measuring system takes no current from the circuit being examined.

Using a voltage calibrator box

A similar zero current measuring scheme to the potentiometric method uses a decade voltage calibrator instead of the potentiometer as shown in Figure 1.21. The voltage calibrator consists of an accurate voltage generator, usually based on a precision Zener

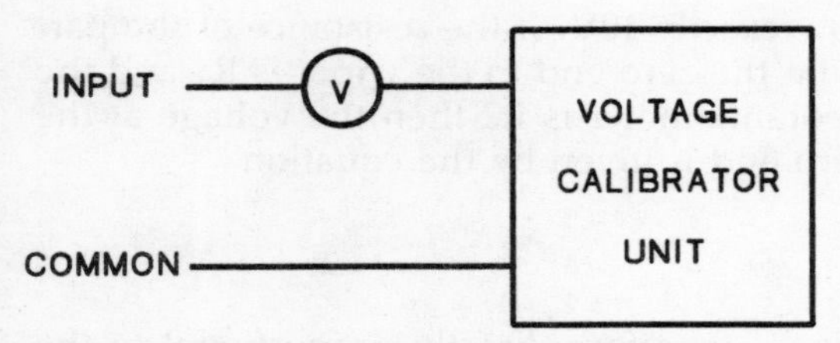

Fig 1.21 Voltage measurement using a voltage calibrator.

diode, combined with a switched decade potentiometer. The setting of the potentiometer is adjusted by means of three or four decade switches so that precise output voltages to perhaps three places of decimals can be selected.

In this scheme there is no need for a switch since the unknown voltage is compared directly with the output of the voltage calibrator unit. The output level of the voltage calibrator is adjusted until the current reading in the meter is zero. At this point the voltage can be read off from the dials of the voltage reference source. Most voltage calibrator units incorporate a centre zero meter for use with this type of measurement.

2 Digital meters

In the last chapter we examined the analogue type meter where the value of the parameter being measured is estimated from the position of a pointer along a calibrated scale. Even when using a high grade laboratory meter of this type it is difficult to take readings with a precision which is better than about 1% of the full scale value. This limitation is largely imposed by the physical arrangement of the scale and pointer scheme. For more precise measurements it would be better if the actual value of voltage or current could be displayed directly as a numerical value.

By using digital logic circuits it is possible to achieve this objective but the circuits required are relatively complex and until the advent of digital integrated circuits this type of instrument would have been impracticable. Today a digital type voltmeter can be built using a single integrated circuit device together with a liquid crystal type display and a handful of discrete components. Meters which provide a digital readout of voltage or current are frequently used for panel displays on electronic equipment and are also used in multi-range test instruments for both laboratory and field use.

Digital logic signals

Before going on to examine the operation of a digital meter it might be as well to look at the way in which signals in a digital logic system operate. In an analogue circuit the voltage or current may be varied in magnitude through an almost infinite number of different levels. As an example if a lamp is connected to a battery through a variable resistor then the brilliance of the lamp can be set to any one of a wide number of brightness levels by adjusting the resistor. The light output is effectively an analogue signal.

A digital logic signal can have only one of two preset levels which are known as the '0' and '1' states. This is equivalent to connecting our lamp to the battery through a switch. Now the lamp is either on ('1' state) when the switch is closed or off ('0' state) when the switch is open. The light output is now a digital signal.

Continuing with the idea of our digital lamp we have the situation that a single lamp could be used to represent the numbers 0 and 1. If we wanted to indicate higher numbers then more lamps would be needed. We could have three lamps which were labelled 1, 2 and 3 and arrange that by switching on one lamp at a time the numbers 0, 1, 2 and 3 could be indicated. This scheme is not very efficient because only one lamp is allowed to be on at any time. A better scheme is to let the first lamp represent the value 0 or 1, the second lamp 0 or 2 and the third lamp 0 or 4. Now there are eight possible combinations of on and off states for the three lamps and if we add together the values for those lamps that are lit we can represent all of the numbers from 0 (all lamps off) to 7 (all lamps on).

In a digital logic system each of the individual logic signals used in combination to represent a number is called a binary digit or bit. The set of bits which make up a numerical value is called a word. In the data word each bit is allocated a weight value in the sequence 1, 2, 4, 8, 16, 32, 64, 128 and so on as shown in Figure 2.1. The value of the data word is obtained by adding together the

Fig 2.1 Bit values in a binary data word.

weights of all of the bits in the word that are set at the logical '1' state. For a word containing 8 data bits the numerical value can range from 0 to 255. Each time an extra bit is added to the word the maximum size of number that can be represented is doubled so that a 12 bit word allows numbers up to 4095 and a 16 bit word gives numbers up to 65535.

In the real world we use the decimal system of numbers where each digit has 10 possible values from 0 to 9 and the digits represent units, tens, hundreds and so on. For any instrument we would obviously like to have the display in decimal form. Conversion from the pure binary type logic signal to decimal form is relatively complex so it is usual to employ a modified way of coding the digital bits to make the conversion process simpler. In this scheme which is known as binary coded decimal or BCD a 4 bit data word is used to represent each of the decimal digits. The 4 bit word is capable of representing the values from 0 to 15 but for the BCD format only those combinations representing the numbers 0 to 9 are used. Separate 4 bit data words are used to represent the units, tens and hundreds digits of the decimal number as shown in Figure 2.2. The main advantage of this form

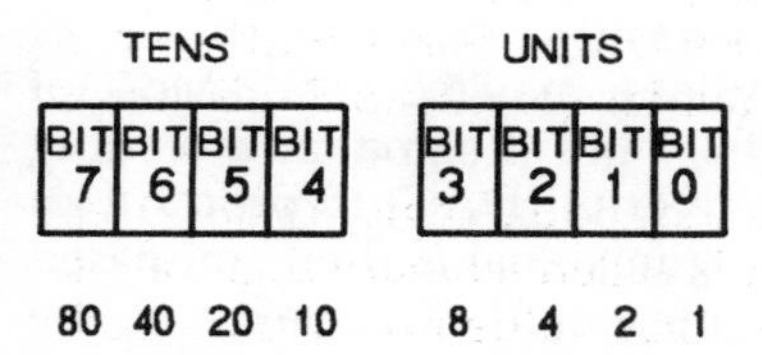

Fig 2.2 The Binary Coded Decimal data format.

of coding is that each set of four bits can be used to drive one digit of the decimal display readout without the need for a binary to decimal conversion.

In typical solid state logic circuits the logical '1' state is represented by a voltage level between about +2.5V and +5V. The logical '0' state is usually represented by a voltage level between 0V and +0.5V.

Basic principles

In the digital voltmeter the input signal is an analogue signal which might have any value between say 0V and 2V. Suppose we wish to have a digital readout which indicates values from 0 to 1.999V. We need some device which can convert the actual voltage level of the input signal into a number which ranges from 0 to 1999. Here we have ignored the decimal point since this can be set in a fixed position on the display to give the appropriate readout. The circuit which performs this function is appropriately called an analogue to digital converter or ADC. Once the analogue value has been converted into a digital number, usually in BCD format, the digital logic signals can be used to drive a decimal display to give a digital readout of the input voltage. If the readout gives values from 0 to 1.999V then it is theoretically possible to measure voltage levels to a precision of some 0.005% of full scale which is considerably better than the result that could be achieved with an analogue type meter.

In most digital test meters the conversion technique involves conversion of the input voltage into a proportional period of time. The time period is then measured by using a digital counter which is fed with an accurate timing clock. For a voltmeter that is to produce a readout of 0 to 1.999V on the display the timing of the counter would be adjusted so that the counter gives a maximum count of 1999 when the input voltage is 1.999V.

For high speed data logging, various alternative types of analogue to digital conversion technique may be used. Most of these involve generating a trial digital output signal and feeding this through a digital to analogue converter (DAC) to produce an analogue reference signal. This analogue signal is then compared with the input signal that is to be measured. The value of the digital output is then modified until eventually the two analogue signals have the same value. At this point the digital output is equivalent to the input analogue signal and the result is transferred to the output display. The advantage of this type of conversion technique is that it is fast and permits virtually instantaneous digital output.

Single ramp voltmeter

Let us start by taking a look at the simplest approach to producing a digital voltmeter. In this type of meter conversion of the input voltage into a time period is based on the way in which the voltage across a capacitor changes with time as it is charged up from a discharged state.

The charge Q contained in a capacitor is given by the equation

$Q = C.V$ coulombs

where C is the capacitance in farads and V is the voltage across the capacitor.

The value of Q can also be expressed as the product of the current flowing in the circuit (i) in amperes and the time elapsed (t) in seconds. Thus we can replace Q to give a new equation

$$i.t = C.V$$

which can be rearranged by dividing both sides by C to give

$$V = \frac{i.t}{C}$$

The value of the capacitance is effectively a constant and if we can arrange that the current i is also held constant then the equation can be simplified to become

$$V = k.t$$

where k = i/C. The voltage across the capacitor is now directly proportional to time and rises linearly with time.

To convert the input voltage into a proportional period of time we can compare the capacitor voltage with the input voltage and measure the time it takes for the capacitor to charge to the same voltage level as the input. The time period can be measured by using a clock generator which produces output pulses at a fixed time interval and then counting the number of pulses that occur whilst the capacitor charges up to the same voltage as the input signal. The counter starts from zero when the capacitor starts to charge and is stopped when the input and capacitor voltages become equal. The number of pulses counted can then be displayed as a decimal number and if the scaling is arranged correctly this will also be a decimal readout of the input voltage..

A simple constant current source for providing the charge current to the capacitor can be produced by using a transistor as shown in Figure 2.3. In a transistor the collector current is almost entirely governed by the value of base to emitter current and will remain virtually constant when the collector to emitter voltage is varied. This assumes that the collector voltage is above about 0.75V. The level of the charge current fed to the capacitor is determined by adjusting the base current to the transistor. As the capacitor charges up the collector to emitter voltage across the transistor falls but the collector current remains virtually constant.

The basic block diagram for a single ramp digital voltmeter is shown in Figure 2.4.

The control of the system is governed by a flip-flop (FF) which

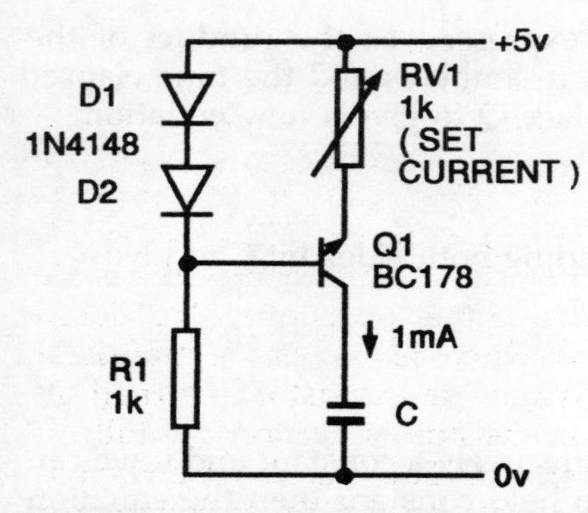

Fig 2.3 Constant current generator using a transistor.

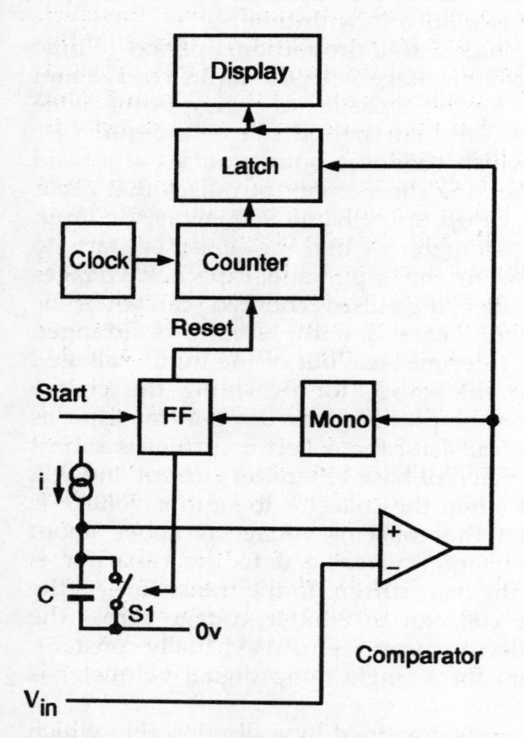

Fig 2.4 Block diagram of single ramp voltmeter.

is in its reset state before the measurement starts. When the flip-flop is reset its output is used to reset the counter to zero and hold the clock gate closed so that no clock pulses are applied to the counter. This ensures that the count will be zero at the start of the measurement cycle.

At the start of a measurement cycle the capacitor must be completely discharged so that V_c starts off from zero. This could be achieved by simply short-circuiting the capacitor with a closed switch. In practical circuits the switch is usually a transistor connected across the capacitor. When the transistor is turned off it has no effect but when the transistor is turned on fully it effectively short-circuits the capacitor. The transistor switch is controlled by the flip-flop and is turned on to discharge the capacitor when the flip-flop is in the reset state.

To start the measurement cycle the flip-flop is triggered into its set state by a start pulse which might be generated by a manual push button. When the flip-flop is set it opens the switch across the capacitor, releases the counter reset line and opens the clock gate to allow clock pulses through to the counter. At this point the capacitor will start to charge up and the counter counts off clock pulses. The counter itself is usually designed so that its count value follows the BCD format so that the count can be used directly to give a decimal readout on the display.

During the measurement cycle the capacitor voltage rises linearly with time and is fed to a comparator circuit where it is continuously compared with the input voltage that is to be measured. The comparator itself consists of a high gain operational amplifier which is set for maximum voltage gain. The unknown voltage is applied to the non-inverting input of the amplifier and the capacitor voltage to the inverting input. When the capacitor voltage is lower than the unknown voltage the output of the comparator circuit is at its maximum positive level. When the capacitor voltage rises to the same level as the unknown voltage, the comparator output switches rapidly to its maximum negative state.

The change of the comparator output from positive to negative is used to trigger a monostable circuit which produces a short time delay. Whilst this delay is taking place the contents of the counter are transferred to a series of latch circuits which drive the display. The display will now show the count value at the time when the comparator switched states. At the end of the delay produced by the monostable circuit the flip-flop is reset to its original state which discharges the capacitor and resets the counter to leave the system ready for a new measurement cycle. The latch circuits retain the last reading from the counter and this continues to be displayed although the counter itself will have been reset by the flip-flop.

In a practical circuit the measuring cycle is repeated at frequent intervals to provide a dynamic display of the voltage being measured. This can be achieved by using a second monostable delay which is triggered when the first monostable times out. The second monostable provides sufficient delay to allow the capacitor to be fully discharged and at the end of this delay time the output from the second monostable is used to generate a new start pulse and the measurement cycle begins again.

In this simple approach to digital voltage measurement there are a number of factors which can affect the result obtained. Firstly the value of the capacitor and the level of the charging current will determine the rate at which the capacitor charges. This charging rate determines the scaling factor between the measured time interval and the input voltage.

The value of the capacitor presents a problem with calibration since the actual capacitance of the average capacitor has a tolerance of as much as 20% relative to the nominal marked value. It is possible to obtain capacitors with value tolerances of 5% or better but these are usually expensive. A component of the desired value could be selected by measuring the values of a batch of capacitors. Alternatively the actual value of the component being used could be measured and appropriate corrections could be made to the level of the charge current to give the correct scale factor. Changes in the capacitance value with temperature can be kept to a minimum by using a polyester type capacitor.

The charge current depends upon the characteristics of the transistor used as a constant current source and may vary with temperature and time. This means that the digital voltmeter may need to have its charge current level adjusted periodically so that the digital display indicates the correct result when a precisely known reference voltage is applied at the input.

The frequency of the clock pulses applied to the counter will also determine the scaling between the time interval being measured and the counter output. Any variation in the frequency of the clock pulses will alter the calibration of the meter. This generally rules out the use of simple clock oscillators based on resistor-capacitor timing circuits. For reliable results a quartz crystal controlled oscillator is essential in order to ensure a stable clock frequency.

With careful design the single ramp digital voltmeter can produce quite good results but is not widely used in modern instruments.

Stepped ramp technique

An alternative approach to the single ramp conversion scheme is shown in Figure 2.5. In this arrangement the linear voltage ramp is derived from the counter itself by feeding the counter output reading to a digital to analogue converter (DAC). This device converts the digital number from the counter into a proportional voltage. Since the counter increments at a constant rate the output of the DAC will rise linearly with time. In fact the ramp is not smooth, as it would be with a capacitor circuit, but rises in a series of small steps to produce a staircase type voltage signal.

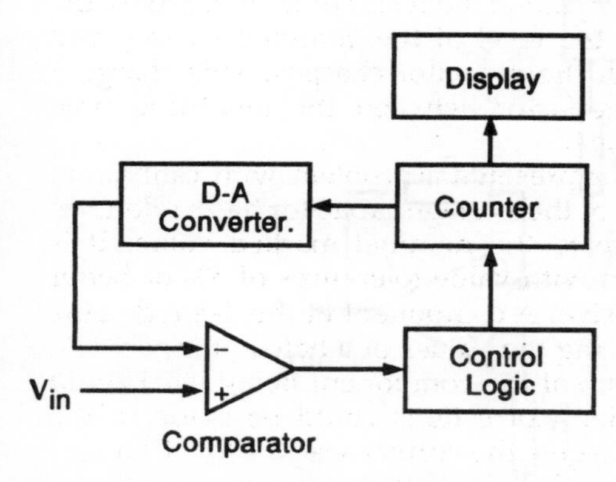

Fig 2.5 Voltmeter using DAC feedback system.

At the start of the conversion the counter is set at zero and the output of the DAC will also be zero. As the count progresses the output voltage from the DAC rises and is compared with the input signal. When the output from the DAC equals the input voltage the comparator switches its output state and as before the count value is transferred from the counter into a series of latch circuits which drive the display. The counter can then be allowed to continue counting. When the counter reaches its maximum count the next clock pulse will set the counter to zero again and a new conversion cycle commences automatically. As the counter returns to zero the DAC output also falls to zero and the comparator switches back to its original state.

This arrangement for a single ramp digital voltmeter overcomes many of the problems presented by the simple capacitor type single ramp circuit. Because the staircase ramp is generated directly from the counter it is automatically locked to the counter clock fequency. Any drift in clock frequency therefore will not alter the calibration and becomes unimportant. If the DAC circuit is properly designed it provides a precisely defined ramp with exactly equal voltage steps at each counter clock period. This eliminates the problems associated with maintaining constant charge current in a capacitor. There is now no capacitor involved in the circuit so variations in capacitance are no longer relevant.

The digital to analogue conversion could be achieved by the arrangement shown in Figure 2.6. Here the four outputs from a

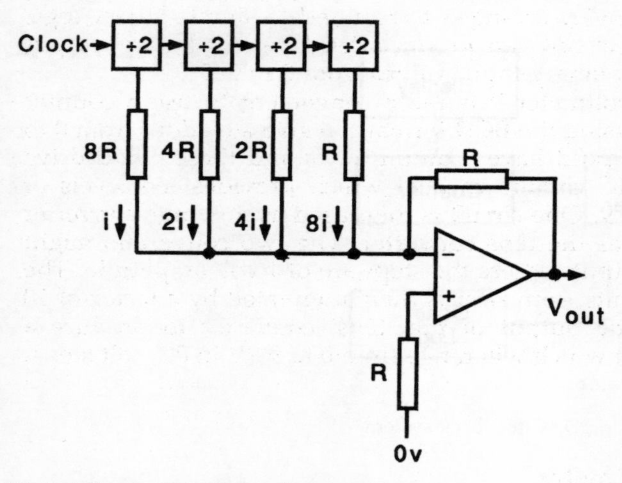

Fig 2.6 Simple digital to analogue converter.

four stage binary counter are fed via resistors to the input of an amplifier. The values of the resistors are weighted so that the first stage of the counter when switched on will pass a current of i milliamperes to the amplifier whilst the second, third and fourth counter stages produce currents of 2i, 4i and 8i respectively. The sum of the current at the input to the amplifier will now range from 0 to 15i and is proportional to the count value held in the counter stages.

If the counter is clocked through its 16 count states the output from the amplifier will be a staircase waveform where the voltage level rises in 16 small steps in sympathy with the count. The size of the voltage steps is governed by the output voltage levels of the counter stages and the gain of the amplifier and may be set to any desired level.

One problem with this simple scheme is that the outputs from the stages of a counter circuit do not switch accurately between zero and some exact voltage so that the currents in the resistors are not precisely defined. In a practical circuit each of the resistors is connected to an electronic switch which is controlled by the counter stage logic output. The switch connects the resistor to ground, when the logic level is 0, or to a fixed reference voltage when the logic level is 1.

The complete DAC circuit can be obtained as a single integrated circuit which might have eight digital input lines and an analogue output. If the DAC is arranged to respond to simple binary logic inputs then the circuit can produce 256 levels of binary output corresponding to binary input values from 0 to 255.

For a digital voltmeter it is more convenient to use a counter which operates using the BCD format. To give a readout from 0 to 99 the counter would have 8 output lines and these could drive two 4 input DAC circuits each of which provides ten levels of output from 0 to 9. One circuit is then used as the units converter and the second as the tens converter. The two converters might each produce outputs where the steps are of 1 volt amplitude. The output of the 'units' converter is then attenuated by a factor of 10 and added to the output of the 'tens' converter to produce a combined output which will range from 0 to 9.9V in 0.1 volt steps.

Dual ramp voltmeter

As we have seen, the main problems with the basic single ramp digital voltmeter are caused by the difficulty in maintaining accurate values for the capacitor, the value of the charging current and the timing of the clock which drives the counter. These variables can be eliminated by using an alternative approach known as the 'dual slope' technique. A block diagram of a meter using this scheme is shown in Figure 2.7.

In the dual slope voltmeter circuit the measurement cycle is divided into two parts. During the first part of the cycle the capacitor is charged up by a current which is proportional to the input signal. This process continues for a fixed period of time during which the counter circuit cycles once through its entire count range. At the end of this time period the capacitor will be

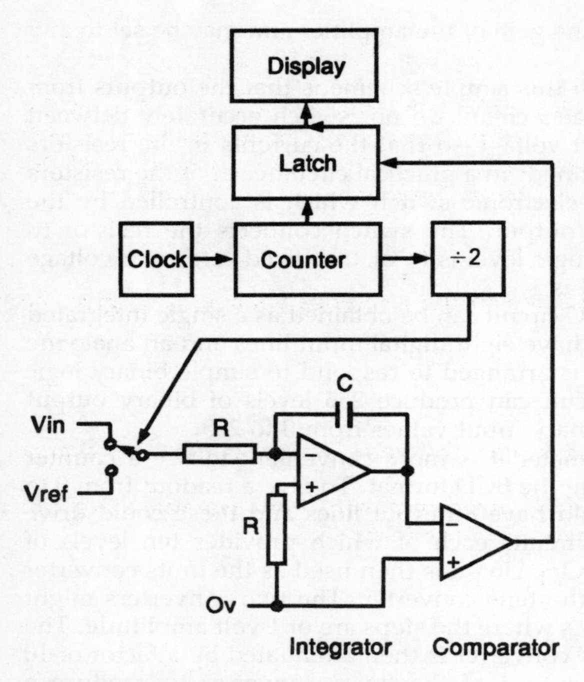

Fig 2.7 Block diagram of dual slope voltmeter.

charged to a voltage which is proportional to the input signal voltage.

For the second part of the measurement cycle the capacitor circuit is switched so that the capacitor is discharged by a current which is proportional to a fixed precision reference voltage. During this discharge period the counter counts up again from zero and the voltage across the capacitor is fed to a comparator which produces an output signal when the capacitor voltage falls to zero.

During the charging period the change in voltage V_c is given by

$$V_c = \frac{I_i \cdot T}{C}$$

where I_i is the charge current produced by the input signal, C is the capacitance of the capacitor and T is the time period for the counter to run through its entire count range. Since the current I_i is proportional to input voltage V_i we can rewrite this equation in terms of V_i as follows

$$V_c = \frac{k \cdot V_i \cdot T}{C}$$

During the discharging period the change in voltage across the capacitor is again V_c since the capacitor is being discharged back to 0V. Now V_c can be related to the discharge current I_r and the discharge time period t by the equation

$$V_c = \frac{I_r \cdot t}{C}$$

Again this can be related to the reference voltage V_r to give the equation

$$V_c = \frac{k \cdot V_r \cdot t}{C}$$

where the constant k has the same value as for the charge equation.

Since both C and V_c are common to both equations we can rewrite the two equations in the form

$$k \cdot V_i \cdot T = k \cdot V_r \cdot t$$

or

$$\frac{V_i}{V_r} = \frac{t}{T}$$

Since T is a constant the count produced at the end of the discharge time t is proportional to the ratio of V_i to V_r.

In this scheme the actual value of the capacitor does not directly affect the results since the same capacitor is used during both charge and discharge cycles. The same reasoning applies to the clock frequency provided that this remains essentially constant during the period of the measuring cycle.

The capacitor charging circuit usually takes the form shown in Figure 2.8 where the capacitor is connected between the output and the inverting input of an operational amplifier. This type of circuit is known as an integrator. The charging current for the capacitor is determined by the series input resistor R and since the amplifier input terminal remains virtually at zero volts the charge current for the capacitor is given by the voltage across the resistor divided by the resistance value. The output from the amplifier is a linear ramp which goes negative when its input voltage is positive and vice-versa. An inverting buffer amplifier is added to correct for this signal inversion. A solid state change-over switch

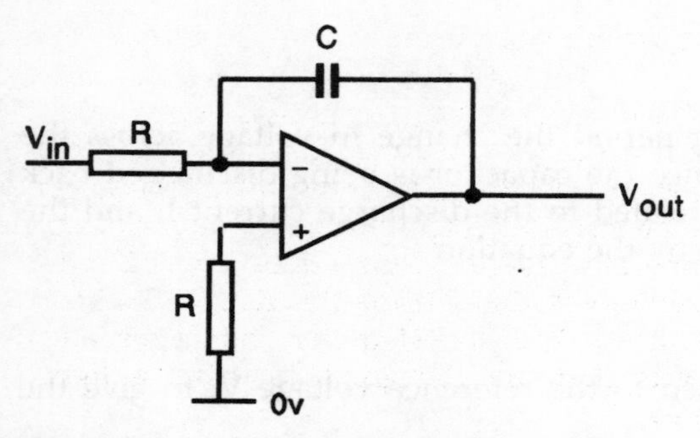

Fig 2.8 Integrator circuit for dual slope meter.

at the buffer amplifier input is used to connect either the input signal to be measured or the reference signal according to the part of the measurement cycle being carried out. Thus for the first part of the measurement cycle the input being measured is applied until the counter reaches its maximum count then the switch is changed over to apply the reference voltage. The reference voltage is arranged to have the opposite polarity to the input voltage so that it will discharge the capacitor back to zero volts.

In a typical circuit the contents of the counter are copied to a set of output latches at the end of the discharge period when V_c has reached zero volts. At this point the two ends of the capacitor are clamped to zero volts by transistor switches. The counter is allowed to continue until it reaches its maximum count and then on the next clock pulse a new measurement cycle is started. At this point the clamp is removed from the integrator circuit and the capacitor starts to charge from the input signal again.

Speed of conversion

The speed of conversion is governed by the clock frequency for the counter and the maximum count provided. For a converter using a 3½ digit display with a count from 0000 to 1999 the counter will count off 2000 clock pulses whilst the capacitor is being charged and a further 2000 clock pulses during the second part of the measurement cycle. The total conversion time is therefore 4000 clock periods. Thus if the clock frequency is 12 kHz the conversion time would be approximately ⅓ second and this is a fairly common value for the conversion time in commercial digital voltmeters.

ICL7106/7107 digital voltmeter chips

A typical integrated circuit for use in producing a dual slope digital voltmeter is the GE/Intersil ICL7106. This chip includes a built in reference voltage generator and clock oscillator as well as the basic dual slope digital voltmeter circuit. Output signals are provided which are suitable for operating a 3½ digit liquid crystal type display. An alternative version of this chip is the ICL7107 which has similar internal circuits for the digital meter, reference and clock but provides outputs that are designed to drive a light emitting diode type display.

The pin connections and signal functions for the 7106 are shown in Figure 2.9 and those of the 7107 in Figure 2.10. By adding a few resistors and capacitors either of these chips can be used to construct a simple single range digital panel meter capable of measuring DC voltages from −1.999V to +1.999V. By altering the values of a few of the components the meter circuit can be adjusted for higher sensitivity when its voltage range becomes −199.9 mV to +199.9 mV.

Both chips include an automatic zero level correction circuit. This uses an additional capacitor which is connected in series with the inverting input of the integrator circuit. When the integrator is clamped at the end of the measuring cycle this capacitor becomes charged by any amplifier offset voltages in the integrator and

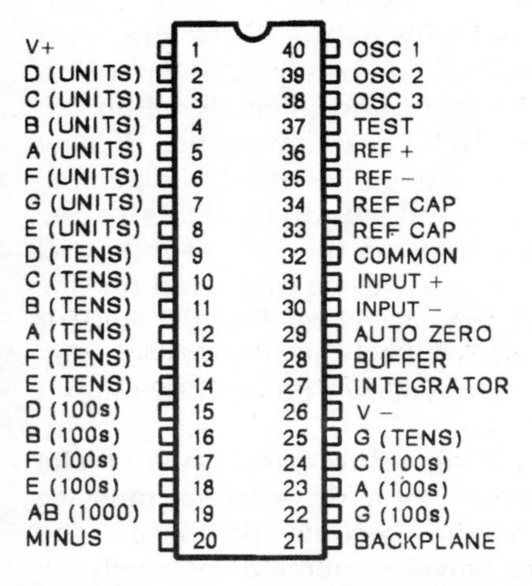

Fig 2.9 Pin connections of ICL7106 DVM chip.

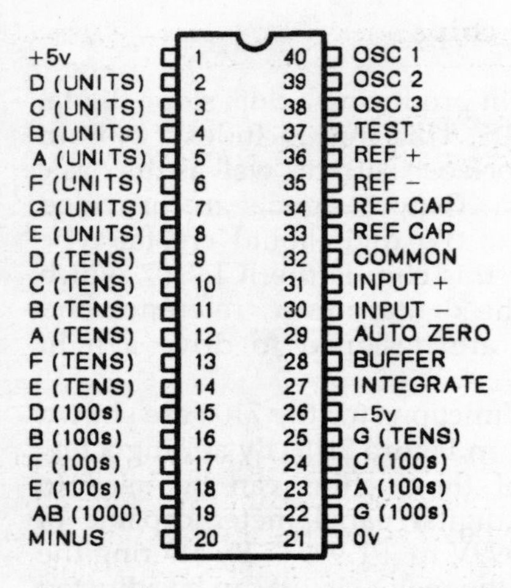

Fig 2.10 Pin connections of ICL7107 DVM chip.

comparator circuits. During the next measurement cycle this offset error voltage is effectively subtracted from the voltage on the integrator capacitor and corrects out the offset errors.

The 7107 chip provides current limited outputs to drive a set of four seven segment led type displays. These displays should be of the common anode type and the anode pins of all four digits should be tied to the +5V supply rail. The most significant digit of the display uses segments A and B to produce a '1' digit and the cathodes for these two segments should be wired in parallel to pin 19 on the 7107 chip. The polarity or minus sign signal on pin 20 of the 7107 may be used to drive the G cathode of this display digit to produce a minus sign when a negative input is detected. Cathodes A to G of the units, tens and hundreds digits of the display are wired to the appropriate pins on the 7107 as indicated in Figure 2.10. A decimal point will need to be displayed on one of the digits of the display to give the correct reading for volts or millivolts. The decimal point cathode of the appropriate display digit is simply connected to the zero volt supply line through a 180Ω current limiting resistor.

The 7106 version of the chip is designed to drive liquid crystal type displays. In this type of display an ac drive signal must be used otherwise the elements would become polarised. The segment drives are in fact square waves generated internally by the chip and a further square wave signal to drive the backplane of the display is output on pin 21. When a segment is to be off, its drive signal is in phase with the backplane signal so there is no voltage difference between the segment and the backplane. When the segment is to be on its signal switches 180 degrees out of phase with the backplane signal. Thus on one half cycle the segment is 5V positive to the backplane and on the next half cycle it becomes 5V negative relative to the backplane.

One problem is that the 7106 does not provide a drive signal for the decimal point element on the display. In order to illuminate this element of the display a signal can be derived from the backplane signal which is inverted before being applied to one of the decimal point elements of the display. The inversion can readily be achieved by using a gate as shown in Figure 2.11. If a

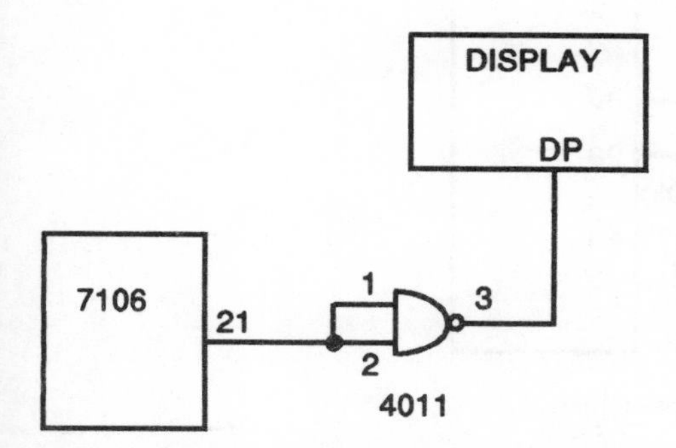

Fig 2.11 Decimal point drive circuit for 7106.

four digit liquid crystal display module is used this may have a special segment for the minus sign and this should be driven from the signal on pin 20 of the 7106.

Simple 2 volt digital panel meter

The basic circuit arrangement for simple digital panel meter with a range of −1.999V to +1.999V using either the 7106 or 7107 is shown in Figure 2.12. To simplify the diagram the connections to the display are not shown. It should be noted that the 7107 version requires a +5V and a −5V supply with the common zero volt line connected to pin 21. The 7106 does not require a dual supply so the V+ and V− pins can be fed from 9V battery. On this chip pin

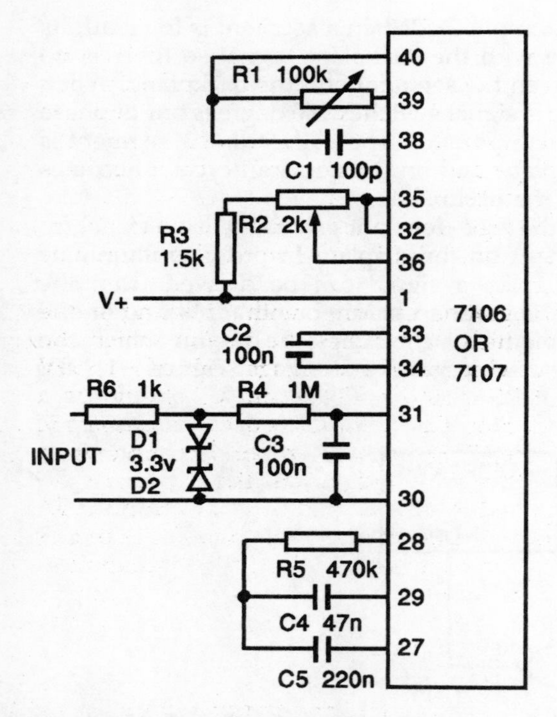

Fig 2.12 Simple 2 volt digital panel meter.

21 is used to output the backplane drive signal for the liquid crystal display and must not be used as a ground pin.

The frequency of the internal oscillator is governed by R1 and C1. The variable resistor R1 should be adjusted to produce an oscillator frequency of about 50 kHz which gives a conversion rate of about three measurements per second. If desired R1 could be made up of an 82kΩ fixed resistor in series with a 22kΩ variable resistor. Setting the variable resistor to about the middle of its scale should give approximately the correct frequency.

To give a full scale range of −2V to +2V on the display a reference voltage of 1V is required between the reference inputs (pins 35 and 36). It should be noted that the two reference voltage inputs are floating so the actual DC level is not critical provided the voltage difference between the pins is accurate. Capacitor C2 acts as a decoupling capacitor for the internal reference circuits.

The voltage level on pin 32 is internally stabilised so that it is 2.8V below the positive supply rail and this voltage may be used to provide a reference signal. Reference input pin 35 is tied to pin 32 and a portion of the 2.8V difference between pins 32 and 1 is fed to the other reference input on pin 36. Potentiometer R2 should be a multi-turn cermet type and is adjusted to produce a voltage of exactly 1V between pins 35 and 36. The setting of R2 could be made by applying the voltage from a single fresh dry cell to the voltmeter input and then adjusting R2 until the display reads 1.54 volts.

Capacitor C5 is the integration capacitor with R5 acting as the current control resistor which determines the rate at which the capacitor C4 charges or discharges. For a voltage scale of −1.999V to +1.999V the value of R5 must be 470kΩ and C5 should be a 220nF polyester type capacitor. Capacitor C4 is used for automatic zero correction and must be 47nF to ensure proper operation when the meter is operating with a 2V full scale sensitivity.

The input signal is applied to pins 30 and 31 of the chip via R4 and C3 which act as a low pass input filter. The two Zener diodes are included to provide protection against the accidental application of an excessive voltage input.

Simple multi-range meter using the 7106/7107

Although the fixed range meter can be useful as a panel display if the meter is to be used as a test instrument it should have several voltage ranges so that it can be used to measure a wide range of voltages from a few millivolts up to hundreds of volts. Figure 2.13 shows the basic circuit arrangement for a multi-range voltmeter based on either the 7106 or 7107 digital voltmeter chip. The ranges provided are 200mV, 2V, 20V and 200V. On each range the meter will automatically read voltages of either polarity.

In this circuit the basic digital voltmeter has been set up to give a full scale sensitivity of −199.9mV to +199.9mV. The reference voltage in this case is set to a level of 100mV but is still derived from the internal stabilised voltage produced at pin 32. For the higher sensitivity the value of the current control resistor R5 is reduced to 47kΩ so that C5 charges and discharges more rapidly. The auto zero capacitor also needs to be changed to 470nF for proper operation of the auto zero function. The display circuits are basically the same as for the fixed range panel meter except that

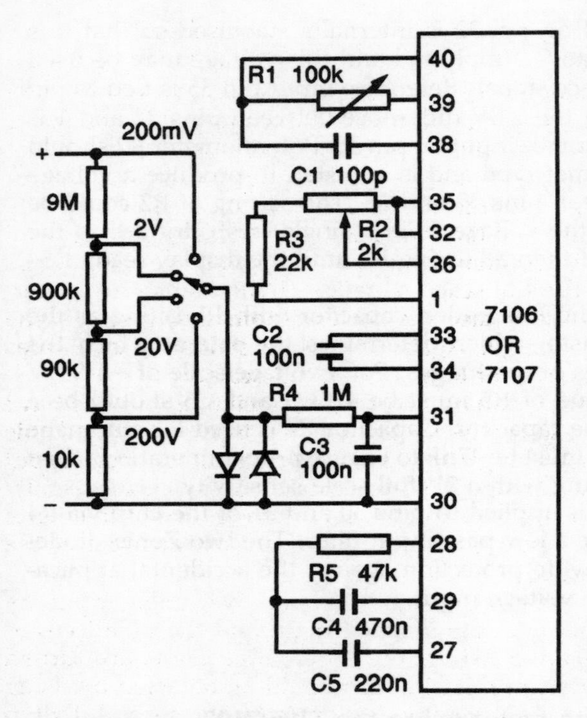

Fig 2.13 Multi-range voltmeter using 7106/7107 IC.

the decimal point drive should be switched to appropriate points in the display by another bank on the range selector switch.

For the 200mV range the input signal is applied via the low pass filter to the input terminals of the voltmeter chip. For the higher voltage ranges an attenuator is switched into the circuit. This takes the form of a potentiometer chain. On the 2V range the input signal needs to be attenuated by a factor of 10 and this is achieved by taking the voltmeter input from the first tap down the potentiometer chain. For 20V the division ratio is increased to 100:1 and for the 200V range the ratio becomes 1000:1. One problem with adding the attenuator circuit is that the input impedance of the meter is reduced to 10 MΩ compared with tens of megohms for the single range voltmeter circuit described earlier. Two silicon diodes are connected across the chip inputs to provide protection in case high voltages are accidentally applied with the meter on a low voltage range.

Auto-ranging

Many of the commercially available digital voltmeters include facilities for automatic range selection. To achieve this the meter circuit detects an over-range condition when the signal being measured is greater than full scale for the range currently selected. When an over-range condition is detected the meter automatically selects the next higher voltage range. This process continues until the meter reading is less than full scale. If the meter reading is less than one tenth of the full scale value the circuit signals an under-range condition. In this case the meter automatically selects the next lower voltage range. Again the range switching continues until the reading is greater than 10% of full scale.

If the meter is fitted with an autoranging facility it will also have input protection so that when a large signal input is applied this will be limited before being applied to the actual meter circuit input in order to prevent damage to the internal circuits. This is usually achieved by having two back to back zener diodes across the meter input and a series limiting resistor to prevent overloading of the circuit being measured.

The 7106 and 7107 devices indicate an overrange input by displaying only the most significant 1 digit, and the polarity sign if the signal is negative, whilst the other three digits are turned off. On the 7107 version this situation could be detected by using suitable logic on the display cathode signals for the second digit to check if all cathodes are at high level indicating that the digit is turned off. The logic signal could then be used to provide automatic range switching to the next higher range. An under-range signal could also be detected because when the signal is below 10% of full scale conditions the first digit will be off and the second digit will be either a 0 or a 1. This condition could also be detected by suitable logic on the display digit drive lines and used to switch the meter to the next lower range. On the 7106 chip the display drive signals are square waves and the logic for detecting over or under range conditions would be more complex.

Noise rejection

In a dual slope integrating digital meter the input signal governs the charging current of the integration capacitor during the initial stage of the measurment cycle. The final voltage to which the capacitor becomes charged will be proportional to the average value of the input signal during the charging period, and any rapid fluctuations caused by noise on the input will tend to be cancelled out. This feature makes the meter relatively insensitive to noise or rapid fluctuations of the signal during the measurement period.

If the meter is powered from the mains there is a possibility that some signal at mains frequency may be injected into the input circuits. One cause for this problem is series injection where leakage occurs back to the mains ground point and a small voltage is developed which is effectively in series with the input signal. Pick up of mains frequency signals on the input leads is usually dealt with by the common mode signal rejection of the input amplifier which has a differential input circuit.

If the clock frequency of the counter circuit is adjusted so that the integration cycle period is a multiple of the period of the supply mains then the integrating action of the meter can be used to remove the effect of any mains signal injected at the input. Since the mains signal is a sine wave and therefore symmetrical about zero its average value over an integral number of cycles is zero so therefore any mains component tends to cancel itself out in the integration process. Typical circuits of digital multimeters usually provide about three conversions per second giving an average voltage taken over 16 cycles (for 50 Hz mains supplies) or 20 cycles (for 60 Hz supplies).

Current measurement using a DVM

Since the digital voltmeter scheme is inherently a voltage measuring device, its circuit needs to be modified in order to measure current. For current measurement the meter must present a low resistance when inserted into the circuit being measured so that the voltage drop across the meter is minimised. The basic technique used is to insert a low value sensing resistor into the circuit where the current is to be measured. The current will develop a small voltage drop across the sensing resistor and this voltage is then measured by the digital voltmeter which should be set for its highest sensitivity. If the series resistor has a value of 1 Ω then a current of 1mA will give a voltage of 1mV. If the digital meter has a scale reading 0 to 199.9 mV then the display reading can be interpreted as a current reading of 0 to 199.9 mA. If the sensing resistor is reduced to 0.1Ω the meter will have a scale of 0 to 1.999A.

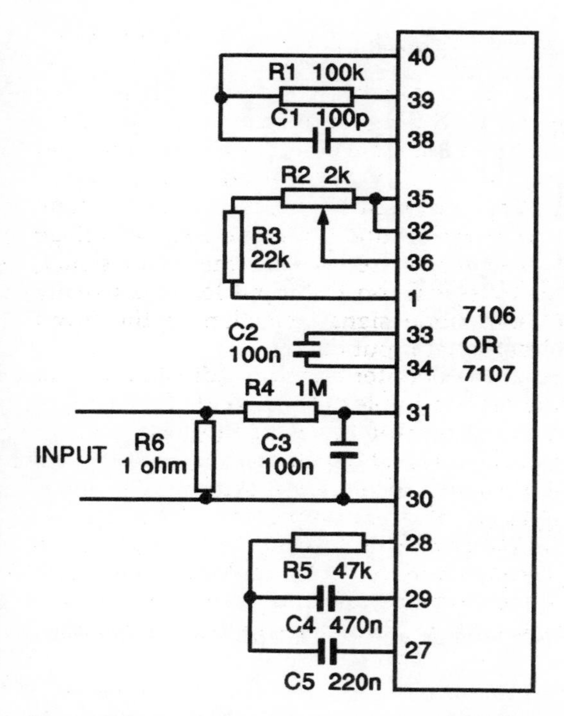

Fig 2.14 Current measurement using the 7106/7 DVM.

When a multi-range digital meter is switched for use as a current meter the low resistance sensing resistor is connected directly across the input terminals of the voltmeter circuit and effectively across the test leads as shown in Figure 2.14. A number of different values of resistance may be switched in to give several current ranges.

AC measurements

Measurment of AC voltages requires the use of a rectifier to convert the input voltage to DC. The rectification could be achieved by using a simple half wave diode peak detector circuit to give a DC output approximately equal to the peak voltage of the input signal. The output of the peak detector is then scaled down

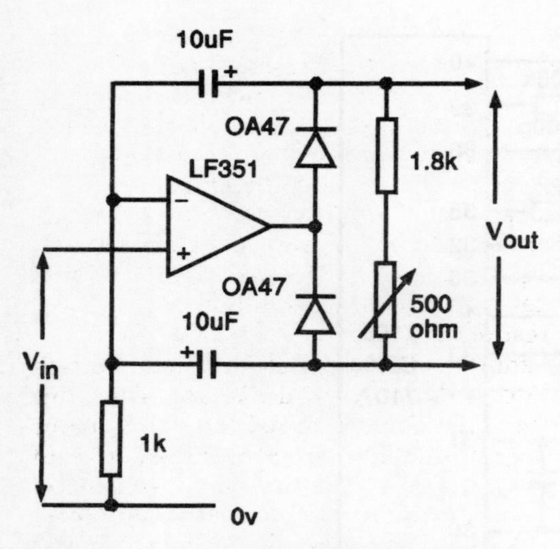

Fig 2.15 Precision rectifier for AC measurements.

to give the equivalent RMS voltage and this is applied to the input of the digital voltmeter.

Most digital voltmeters use a precision rectifier circuit for AC signal measurement. This circuit uses an operational amplifier and two signal diodes as a full wave detector. A typical circuit is shown in Figure 2.15. The output of this circuit is suitable for driving a digital meter with a balanced input such as a meter based on the ICL7106 integrated circuit. The preset variable resistor is used to adjust the scale factor to give the correct RMS reading on a sine wave signal.

In the multi-range meter, shown in Figure 2.13, the precision rectifier may be inserted into the circuit between the attenuator switch and the low pass input filter when AC voltage or current ranges are selected.

Other digital meter chips

For a more advanced digital meter giving 4½ digit readout the GE type ICL7129 chip is available. This, like the 7106/7 type, comes in a 40 pin DIL package. In order to accommodate the additional display digit and other features this device is designed to drive a triplexed LCD type display. In this type of display each digit has its segments divided into three groups with three elements in each group. These comprise the segments a to g, the decimal point and various annunciator messages. There are three backplane elements on the display, each of which activates one group of elements in the display and the three groups are activated in sequence.

This chip provides over-range and under-range output signals which can be used to provide automatic range switching. When set up for a full scale voltage range of −199.99 mV to +199.99 mV this chip produces a digital meter with a resolution of 10μV for each step on the display readout.

The Ferranti ZN451E is another digital panel meter chip which has outputs to drive a 3½ digit liquid crystal display. This chip includes a built in reference voltage and internal timing oscillator and runs from a single +5V supply. The circuit includes an auto zero function which operates digitally and can be controlled by external circuits. This allows offset errors in the external amplifier circuits to be corrected so that it is possible to increase the sensitivity of the meter to produce a full scale reading for an input of only 1.999mV.

Commercial digital meters often use custom designed integrated circuits which permit the addition of other functions, such as transistor tests, capacitance measurement and even frequency measurement within a single instrument. Some instruments use a microprocessor to provide all of the control functions and some of the measurement logic. The advantage of using a microprocessor controlled instrument is that its operating mode and the measurement range can readily be selected by using remote commands, and the readings from the meter can be output as data to another instrument or to a computer for further processing or analysis.

Analogue bargraph readouts

One advantage of the simple analogue meter over a digital meter is that the analogue type can display a slowly varying signal in a readable form. If for instance the signal being examined is varying at a rate of perhaps one cycle every five seconds the needle of an analogue meter will swing slowly back and forth between upper and lower limits and it is possible to visually estimate the limits of the swing and the mean value of the signal. The digital meter on the other hand displays a sequence of rapidly changing numbers which will be difficult to read and any assessment of the signal variation is virtually impossible.

To overcome this disadvantage of the digital readout scheme many of the more expensive digital multimeters have, in addition to the digital readout, a bargraph type display which provides an analogue readout of the signal level. This type of display operates in a similar fashion to a mercury type thermometer where instead of a needle the display consists of a variable length column with a scale mounted alongside. The position of the top of the column indicates the current reading. In an electronic bargraph display the bar consists of a column of individual led indicators or individual segments of an lcd type display. When no signal is present none of the segments is illuminated. As the signal increases the segments of the bar are progressively lit to form a bar whose length is proportional to the level of the signal applied. A special bargraph decoder and driver circuit is used to convert the signal from a binary or bcd type digital format into the drive signal pattern required to produce the variable length bar.

Bargraph type displays have for some time been used as level indicators in audio systems but for this purpose the display is a relatively crude type with perhaps 10 to 12 steps in the bar length. Whilst this is adequate for the subjective indication required on an audio system a much better resolution is required for use in a meter and this should preferably achieve a similar level of precision to that of a standard analogue meter. In most cases this means that the bargraph will have some 40 or 50 segments to give a readout precision of the order 2% of full scale. The bargraph is displayed alongside a calibrated analogue scale which may either be printed on to the display panel for led type displays or incorporated as part of the display itself in a liquid crystal type display unit.

The bargraph data could be derived from the signals which drive the digital readout for the meter and then passed through an appropriate decoder and driver circuit to provide the drive for the bargraph display elements. A better approach is to use a separate analogue to digital converter to drive the bargraph display. With this arrangement the converter driving the bargraph might be set to give 20 samples per second so that the bargraph has a much faster response than the integrating digital meter. This gives the bargraph display a dynamic performance similar to that obtained with a conventional analogue meter.

3 Measuring R, C and L

So far we have looked at how meters can be used to measure voltage, current and power. Another important area of measurement is that of finding the values of resistance, capacitance and inductance of components or circuits. We shall start by looking at techniques for measuring resistance.

Continuity testers

The simplest form of resistance measurement is a continuity test which merely checks to see if there is a conducting path between two points in a circuit. This test simply indicates whether the resistance between the two points is high or low and is convenient for tracing individual wires through a multiwire cable or for tracing out track connections on a printed circuit board.

One popular circuit for a continuity tester is shown in Figure 3.1. Here a buzzer is connected in series with a battery and the two test leads. One test probe is connected to one end of the wire or circuit that is to be checked and the second probe is applied to the

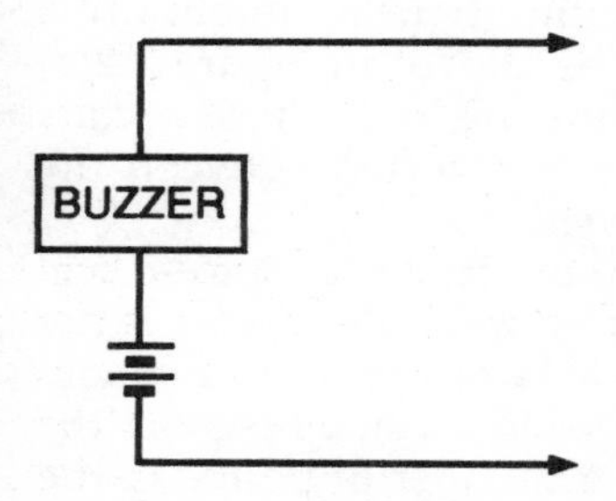

Fig 3.1 Continuity tester using a buzzer.

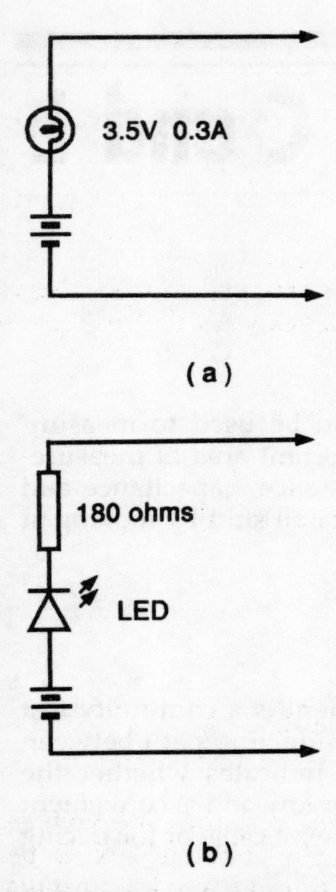

Fig 3.2 Continuity testers using lamps.

other end of the circuit. If the resistance between the two test points is low the buzzer sounds to indicate continuity.

As an alternative to the buzzer the continuity tester might use a filament lamp as a continuity indicator as shown in Figure 3.2(a) or a light emitting diode as shown in Figure 3.2(b). In these circuits the lamp will light up when continuity is detected between the points to which the test probes are applied.

Some care is needed when using a continuity tester since it will still indicate continuity when the circuit between the points is not zero but perhaps a few ohms or even tens of ohms. In a circuit which contains inductors, transformers or low value resistors the tester may indicate continuity between a number of points in the circuit although the actual paths between those points contain resistors or inductors. This can present problems when the tester is used for tracing the path of tracks on a printed circuit board or checking for continuity along a particular track.

Measuring resistance

To measure the value of a resistor one simple approach is to use the circuit shown in Figure 3.3. Here a voltage is applied across the resistor to be tested and a meter in series with the resistor is

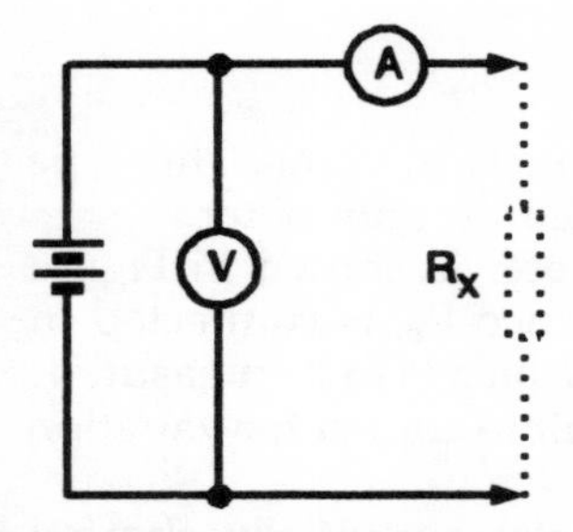

Fig 3.3 Resistance test using ammeter/voltmeter method.

used to measure the current flowing in the resistor. The voltage across the circuit is also measured using a voltmeter. Now by Ohm's law the value of the resistor can be found using the formula

$$R = \frac{V}{I} \text{ ohms}$$

where V is the voltage across the resistor in volts and I is the current through the resistor in amperes. If the current is measured in milliamperes then the resistance value will be given in kilohms.

The ohmmeter

The technique for measuring resistance using a voltmeter and an ammeter is interesting as an educational experiment but is not particularly suited for everyday use since it requires two meters and some calculation in order to find the value of the unknown resistance. There are simpler techniques using a single meter in

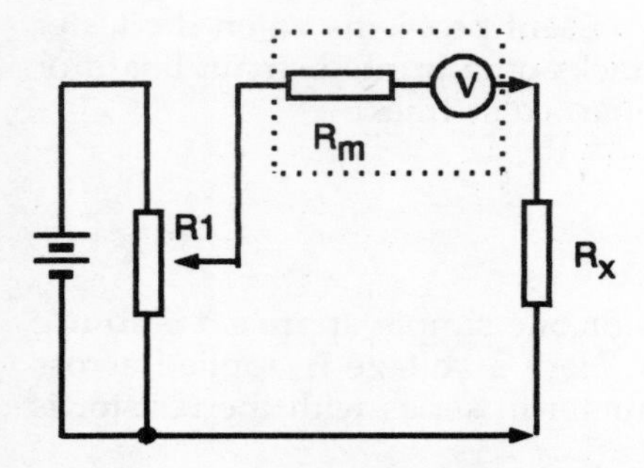

Fig 3.4 Simple ohmmeter for high resistances.

which the meter scale can be calibrated directly in ohms. This type of resistance measuring instrument is called an ohmmeter.

One simple arrangement for an ohmmeter is shown in Figure 3.4. Here a voltmeter with internal resistance R_m is connected in series with a battery and the resistance R_x which is to be measured. A potentiometer R1 is included in the circuit to correct for variation in the battery voltage.

To use this ohmmeter the test terminals are initially shorted together and potentiometer R1 is adjusted so that the meter reads full scale. The resistor to be measured is then inserted between the test terminals which will cause the current flowing in the circuit to fall and the voltmeter reading will also fall. If the unknown resistance equals the resistance of the voltmeter the voltage reading will fall to half scale.

The unknown resistance and the voltmeter resistance effectively form a potential divider across the supply voltage from the potentiometer R1 and the meter reading is given by the equation

$$V_m = \frac{R_m}{R_m + R_x} \cdot V_s \text{ volts}$$

where V_s is the supply voltage, V_m is the meter reading, R_x is the unknown resistance and R_m is the resistance of the meter. It is assumed here that the resistance of the battery and the potentiometer R1 are low compared with R_m.

The resistance scale on this type of meter follows a reverse reciprocal law with the lowest values at full scale and the highest values at the zero end of the meter scale. The resultant scale is very nonlinear with the markings cramped together at both the high and low ends of the scale. This type of circuit works quite well for

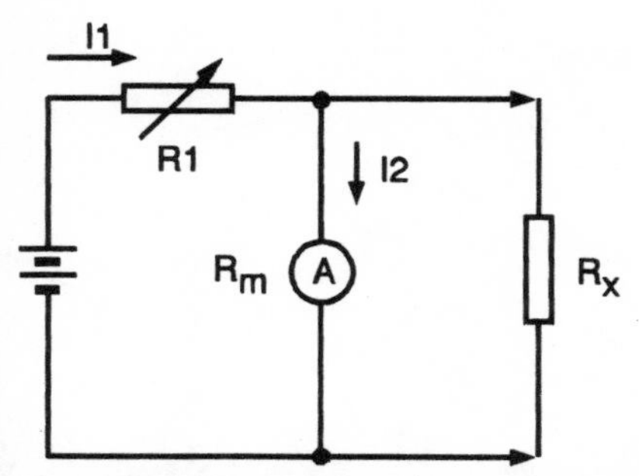

Fig 3.5 Simple ohmmeter for low resistances.

high values of unknown resistance in the range 0.02 R_m to 10 R_m. Thus a voltmeter with a resistance of 1000Ω would work quite well for unknown resistances between about 20Ω and 10kΩ.

For measuring low values of resistance an alternative circuit can be used as shown in Figure 3.5. This is similar to the circuit described in Chapter 1 for measuring the coil resistance of a meter. A battery is used to energise the circuit and variable resistor R1 is used to adjust the current through the meter to give a full scale reading with no external resistance connected to the test terminals. When the resistance to be measured is connected across the terminals it diverts some of the current away from the meter to give a reduced meter reading. If the unknown resistance is equal to the meter coil resistance R_m then the meter reading will fall to half scale. Once again the meter scale can be calibrated directly in terms of resistance.

The unknown resistance R_x is given by the equation

$$R_x = \frac{I2 \, . \, R_m}{I1 - I2}$$

This equation also gives a very nonlinear scale but here the highest values of resistance are at the full scale end and the low resistance readings are at the zero end of the scale. The circuit assumes that the total current flowing through R1 remains constant and this will be essentially true of the value of R1 at least 50 times the meter coil resistance. A typical arrangement might use a 1mA meter with a coil resistance of 50Ω, a 3V battery and a maximum value for R1 of about 3.3kΩ.

The disadvantage of these two simple ohmmeter circuits is that neither can measure a range of resistance values of more than

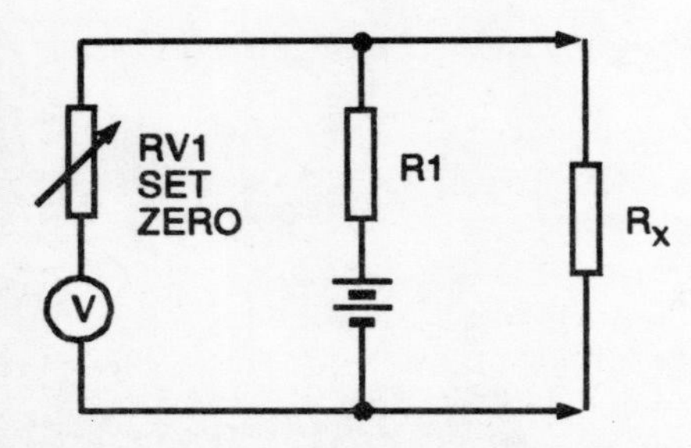

Fig 3.6 General purpose ohmmeter circuit.

about 500:1 with any degree of accuracy. For a practical ohmmeter we need to be able to measure any resistance from perhaps a few ohms to a few megohms. This can be achieved by using the circuit arrangement shown in Figure 3.6.

In this circuit a known resistance R1 is connected in series between a battery and the test terminals. A high resistance voltmeter is used to measure the voltage developed across the test terminals. For proper operation it is assumed that the voltmeter resistance is at least 20 times the value of R1. A variable resistance RV1 in series with the voltmeter is used to correct for variations in battery voltage.

Initially the test terminals are open circuited and RV1 is adjusted to bring the meter reading to full scale. The unknown resistance R_x is then connected between the test terminals and a new meter reading is obtained which depends on the value of R_x compared with R1. The scale can be calibrated directly in terms of resistance and gives a reverse reciprocal scale. The value of the external resistance is given by the equation

$$R_x = \frac{V1.R1}{V2} - R1$$

where R_x is the unknown resistance, V1 is the open circuit voltage and V2 is the voltage when R_x is connected across the test terminals.

The meter will give a useful resistance reading for values from 0.02R1 up to 10R1. By switching in different values of R1 a number of resistance ranges can be covered to give the possibility of reading values of resistance from 1Ω up to perhaps 10 MΩ. For low resistances R1 might be set at 100Ω to give resistance readings up to about 1000Ω. Other values of R1 of 1kΩ, 10kΩ and 100kΩ could be switched in to give scale multiplicaion factors of x10, x100 and x1000 allowing resistance values up to 1 MΩ to be measured.

For the lower resistance ranges a 3V battery might be used to energise the circuit and the voltmeter would be set for about 3V full scale. For measuring resistances above about 10000Ω the battery voltage is increased to perhaps 15V or 25V. This allows the series resistance of the voltmeter circuit to be increased so that it does not appreciably load the circuit when the value of R1 is increased.

Linear ohmmeter

The main disadvantage of the basic analogue ohmmeter circuit is that the scale is nonlinear and also reversed so that the lowest values of resistance appear at the full scale end of the meter scale. It is possible to devise a circuit for an ohmmeter which gives a linear scale with zero resistance at the left or zero end of the meter scale and the maximum resistance value at the right hand end of the scale. The typical arrangement for a linear ohmmeter is shown in Figure 3.7.

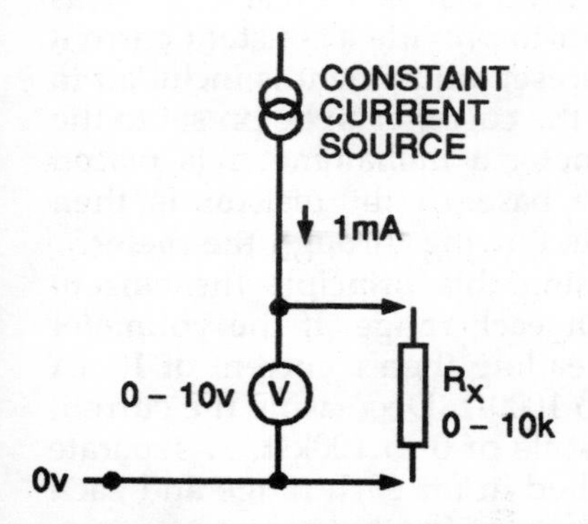

Fig 3.7 Ohmmeter circuit with a linear scale.

If we pass a fixed current of 1mA through the resistor under test the voltage developed across the resistor will be directly proportional to its resistance value. With a current of 1mA the voltage developed is 1V for every 1000Ω of resistance. Thus if the resistor being tested has a value of 2200Ω the voltage across it will be 2.2V.

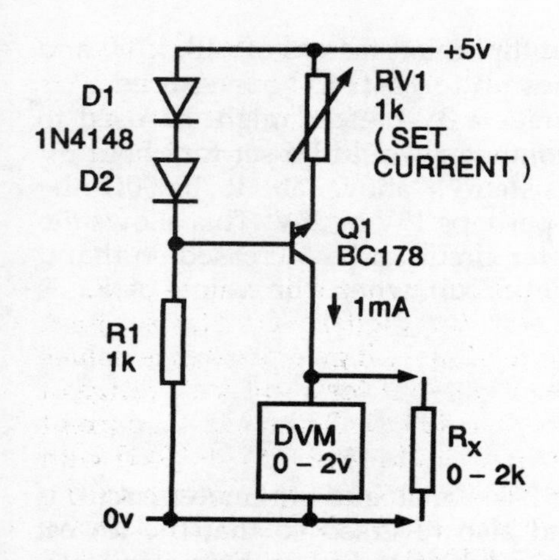

Fig 3.8 Digital ohmmeter using constant current feed.

If a high resistance voltmeter is connected across the test terminals and set to 10V full scale, the meter will read resistances from 0 to 10kΩ directly.

The current that flows through the external resistor must remain constant irrespective of the voltage drop across the test leads. This can be achieved by using a transistor to provide a constant current source as shown in Figure 3.8. A preset adjustment is included in the emitter of the transistor so that the current can be preset to the required value. To calibrate the meter a milliammeter is placed across the test terminals and the base circuit resistor is then adjusted until the desired current is flowing through the meter.

For a multi-range ohmmeter using this principle the current value required will be different for each range. If the voltmeter used has a 10V full scale voltage reading then a current of 10mA would give a resistance scale of 0 to 1000Ω. Decreasing the current level to 100μA gives a resistance scale of 0 to 100kΩ. A separate preset base resistor could be switched in for each range and each would be set to give the required current for that range.

Digital ohmmeters

There are two basic approaches which can be adopted to turn a digital voltmeter, based on a chip such as the ICL7106, into a direct reading digital resistance meter.

The first approach uses the linear ohmmeter scheme with a constant current source in series with the unknown resistor that is to be measured as shown in Figure 3.8. The digital voltmeter is used to measure the voltage drop across the resistor which is given by the equation

$$V = I.R$$

where I is the current and R is the resistance being measured.

If the current is set at 1mA with the digital voltmeter set for a 200mV full scale range then the readout will give resistance values from 0 to 199.9Ω. If the voltmeter were set for a full scale range of 2V the resistance scale would be 0 to 1999Ω. Reducing the current level through the resistor to 100μA would give 0 to 19.99kΩ with the meter set for 2V full scale. A current of 10μA increases the resistance range to give a maximum of 199.9kΩ and 1μA allows resistance measurements up to 1.999MΩ.

Ratio technique

When discussing the digital voltmeter in Chapter 2 it was seen that the reading on the display is proportional to the ratio between the input voltage V_i and the reference voltage V_r. We can make use of this property of the meter to measure resistance and display the result as a digital number.

Suppose we connect a known resistor R_s in series with the unknown resistance R_x and apply a voltage across the pair of resistors as shown in Figure 3.9. Let us assume the current flowing in the circuit is I. The voltage V_s across R_s is given by

$$V_s = I \, . \, R_s$$

and similarly V_x is given by

$$V_x = I \, . \, R_x$$

so the ratio of voltages is the same as the ratio of the resistances

$$\frac{V_x}{V_s} = \frac{R_x}{R_s}$$

If we apply the voltage V_x to the input terminals of the digital voltmeter and use the voltage V_s as the reference voltage for the meter then the reading on the meter will be proportional to the ratio of the resistors R_x and R_s.

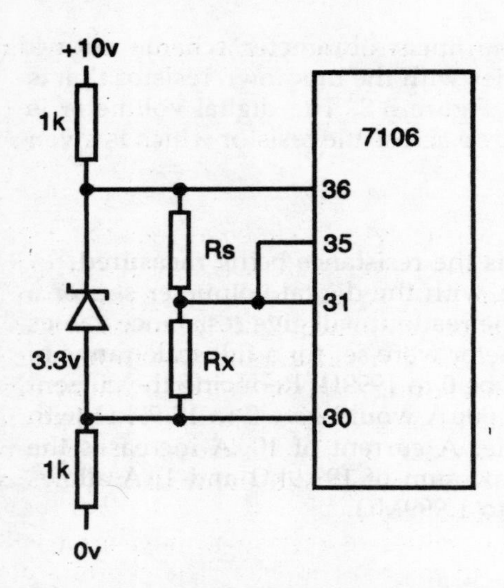

Fig 3.9 Digital ohmmeter using ratio technique.

Suppose we are using a meter based around the ICL7106 integrated circuit. If we make the standard resistance R_s equal to 1000Ω the meter will display the value of resistance R_x as a reading of 0 to 1999Ω. Values of resistance R_x which are 2000Ω and over will produce an overscale indication on the meter.

If the value of the standard resistance R_s is increased to 10kΩ the meter will now indicate resistance values of R_x which range from 0 to 19.99kΩ. By changing the value of R_s we can set the meter to read any desired resistance range. To make the scaling of the display sensible R_s should have a value which is a power of 10 and suitable values of R_s are 10Ω, 100Ω, 1kΩ, 10kΩ, 100kΩ and 1MΩ.

A multi-range digital ohmmeter can now be built quite readily by simply switching in appropriate values of R_s as each range is selected. By using this technique the meter could be made to read resistance within the ranges 0 – 19.99Ω, 0 – 199.9Ω, 0 – 1.999kΩ, 0 – 19.99kΩ, 0 – 199.9kΩ, 0 – 1.999MΩ and 0 – 19.99MΩ.

The advantage of this ratio technique is that the actual value of voltage applied across the pair of resistors is not critical so that there is no need for a series of setting up potentiometers for each range of the meter. The applied voltage does however need to be reasonably stable and can conveniently be derived from the meter circuit power supply by using a resistor and a Zener diode in series.

Insulation testers

For testing the insulation of wiring systems, such as the insulation of mains circuits, the standard instrument is a Megger which applies a high voltage of some 600V to the circuit being tested and measures the effective insulation resistance. Early versions of such instruments generally used a hand cranked generator to produce the high voltage and the measuring part of the instrument has basically the same arrangement as a simple ohmmeter. Modern insulation testers use a transistor oscillator driving a step up transformer to produce the high voltage and the readout may be produced by a built in digital voltmeter.

In an insulation tester the values of resistance measured will range from perhaps 100kΩ up to some 50 MΩ. The centre scale reading might typically be about 5 MΩ so that the meter will have a 5MΩ resistor in series. Typical current when measuring 5MΩ at a voltage of 500V would be 50μA. The series resistor used in the instrument must be able to withstand the high test voltage and might typically be implemented by using a number of resistors in series so that the voltage across each individual resistor is limited to a safe level.

The Wheatstone bridge

A simple analogue ohmmeter is useful for identifying unknown resistors and making approximate measurements of resistance in a circuit. For more accurate measurement of resistance a piece of apparatus known as a Wheatstone bridge may be used.

The basic arrangement of a Wheatstone bridge is shown in Figure 3.10. Here resistors R1 and R2 form a potential divider across the battery terminals and resistors R3 and R4 form a second potential divider circuit. The meter is connected between points A and B and detects any difference in potential between these two points. Resistor R2 is variable and is fitted with a calibrated dial. Resistor R4 is the unknown resistor which is to be measured and is connected across a pair of test terminals. Resistors R1 and R3 are fixed resistors and, for the moment, we will assume that they are equal in value.

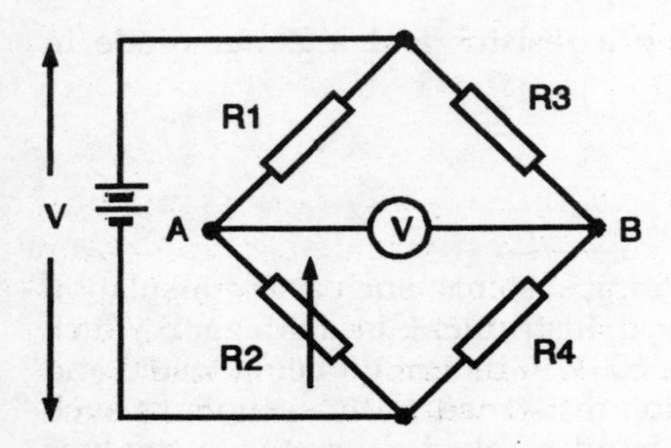

Fig 3.10 The Wheatstone DC bridge circuit.

To make a measurement the unknown resistance R4 is connected to the test terminals on the bridge and the variable resistor R2 is adjusted until the meter reading becomes zero when the bridge is said to be balanced. Since no current flows through the meter the voltage levels at points A and B must be equal.

The voltage at point A relative to battery negative is given by

$$V_A = \frac{V \,.\, R2}{R1 + R2}$$

and the voltage at point B is

$$V_B = \frac{V \,.\, R4}{R3 + R4}$$

where V is the total voltage applied across the bridge by the battery.

When the bridge is balanced the voltages at points A and B are equal so we can combine the two equations to give

$$\frac{R2}{R1 + R2} = \frac{R4}{R3 + R4}$$

which can be re-arranged to give

$$R1 \,.\, R4 + R2 \,.\, R4 = R2 \,.\, R3 + R2 \,.\, R4$$

the R2 . R4 terms cancel out leaving

$$R1 \,.\, R4 = R2 \,.\, R3$$

and the value of the unknown resistor R4 is

$$R4 = R2 \cdot \frac{R3}{R1}$$

If the resistors R1 and R3 are made equal in value then this simplifies to make the unknown resistor R4 equal to the variable resistor R2. If an accurately calibrated linear potentiometer is used for the variable R2 then the value of R4 can be read off directly from the dial fitted to R2. In a typical bridge R2 would be a precision ten turn potentiometer fitted with a calibrated 10 turn dial.

One problem with this simple bridge is that if the value of R4 is larger than the maximum value of R2 it is not possible to balance the bridge since V_A can never be made equal to V_B. A further difficulty is that when the resistance of R4 is small, R2 will be set very close to one end of its range and it becomes difficult to obtain a precise reading from its scale.

In the analysis of the bridge circuit we made the assumption that resistors R1 and R3 are equal in value so that the term R3/R1 on the right hand side of the equation for the value of R4 becomes 1. Suppose we make the value of R3 equal to 10 times that of R1. When this version of the bridge is balanced the equation for the value of R4 becomes

$$R4 = 10\ R2$$

With this arrangement we can measure values of R4 up to ten times the maximum value of R2.

If the value of R1 is made equal to 10 times R3 then the equation for R4 becomes

$$R4 = 0.1\ R2$$

This arrangement is useful because with low values for R4 the balance setting for R2 moves further up its scale to allow a better reading to be obtained from the dial.

The resistors R1 and R3 are normally referred to as the 'ratio arms' of the bridge because they control the scaling factor between the value of the unknown resistor R4 and the setting of the variable resistor R2. For a general purpose bridge the values of R1 and R3 may be switched to give ratios of 0.01, 0.1, 1, 10, and 100 for the resistors R3 and R1. If R2 is a 10kΩ variable it should be possible to measure resistances from a few ohms up to about a megohm.

The indicating meter should be a centre zero type and is set up as a voltmeter with a full scale deflection in either direction equal to the voltage of the battery that is used to energise the bridge. Typically this might be a 50 – 0 – 50μA panel meter with a suitable voltage multiplier resistor connected in series. The DC supply which energises the bridge is usually a battery and might have a voltage of 9V.

As an alternative to the conventional analogue meter it is possible to use a digital panel meter to provide the null indication. The main disadvantage of the digital meter is that, unless the bridge adjustments are made very slowly, it is difficult to assess the change in voltage and the null point could easily be missed. A digital meter with a bargraph readout is more useful for this application since it gives a better indication of the way in which the voltage is changing as the bridge is adjusted. Some digital multimeters allow the bargraph display to be operated in a centre zero mode for use in bridge circuits.

AC bridge circuit

Instead of using a battery or DC power supply to energise the bridge it is possible to use an AC signal. For a general purpose bridge the signal used is generally a sine wave with a frequency of about 1000Hz in the audio frequency range. It is possible to operate a bridge circuit at radio frequencies provided that a suitable detector is used to indicate when the bridge is balanced.

When the bridge is energised by an AC signal the simple DC voltmeter is no longer effective as a balance indicator. One obvious alternative for use as the bridge balance detector is a rectifier type meter. The problem with this solution is that at low signal levels the rectifier meter is insensitive and it becomes difficult to detect an exact null condition.

The usual choice for the balance detector in a bridge operating at audio frequency is a pair of headphones as shown in Figure 3.11. When the bridge is unbalanced an audio frequency tone will be heard in the headphones. As the bridge is adjusted and approaches the balance condition the tone becomes quieter until at the balance point no tone is heard at all. The headphones used for this purpose should preferably be of the high impedance type with a nominal resistance of the order 2000Ω. An alternative scheme is to feed the bridge signal to the inputs of an operational amplifier such as the LM351 and then use the output of the amplifier to drive the headphones. This would allow the use of the more widely available low impedance headphones and will also

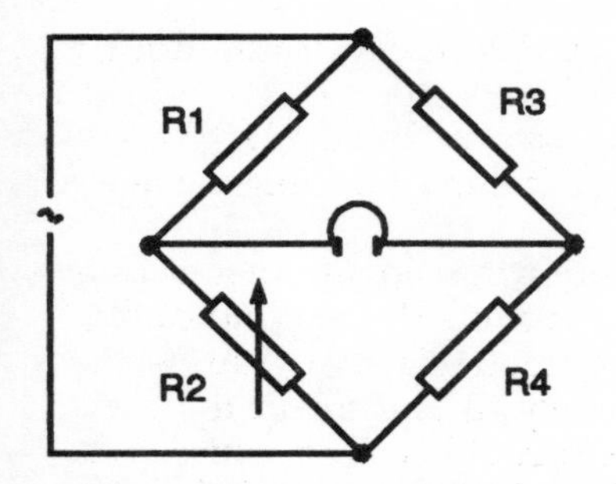

Fig 3.11 Wheatstone bridge using AC energisation.

reduce the loading effects on the bridge to give a sharper null indication. Another way of detecting bridge balance is to feed the bridge output signal to an oscilloscope and then adjust the bridge to give minimum signal amplitude on the oscilloscope display.

If the four arms of the bridge are all purely resistive the results obtained should be exactly the same as for a simple DC energised bridge. There is no particular advantage in using an AC energised bridge for measuring resistance since a DC energised bridge is more convenient for this purpose. The main applications of an AC energised bridge are in making measurements of capacitance, inductance and impedance.

Capacitance bridge

An AC energised bridge can be used to measure capacitance by replacing the two lower arms with capacitor elements as shown in Figure 3.12. In this case C2 is the unknown capacitor and C1 is a calibrated variable capacitor. Resistors R1 and R2 act as the ratio arms for the bridge.

In the capacitor bridge the lower part of each bridge arm presents a capacitive reactance instead of a resistance. The reactance of the capacitor is given by

$$X_c = \frac{10^6}{6.28\,f\,.\,C}\ \text{ohms}$$

where C is the capacitance in microfarads and f is the frequency of the signal applied to the bridge in hertz. Since the frequency

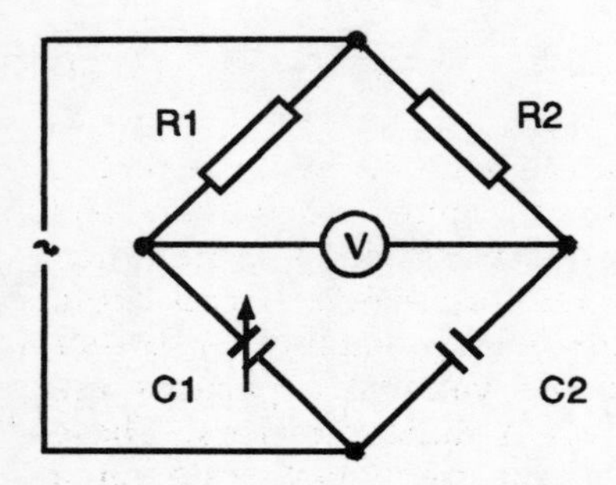

Fig 3.12 Simple capacitance bridge.

at which the bridge is energised is a constant we could rewrite this as

$$X_c = \frac{k}{C}$$

where k is a constant.

If we substitute the reactance terms into the basic DC bridge equation we get

$$X_{c2} = X_{c1} \cdot \frac{R2}{R1}$$

Now by rewriting this in terms of capacitances C1 and C2 we get

$$\frac{k}{C2} = \frac{k}{C1} \cdot \frac{R2}{R1}$$

and re-arranging this gives

$$C1 \cdot R1 = C2 \cdot R2$$

so the value of C2 becomes

$$C2 = C1 \cdot \frac{R1}{R2}$$

This is similar to the equation for a resistance bridge except that the ratio arm term (R1/R2) has been inverted. If we make R1 equal to R2 then the unknown capacitance value can be read off directly from the dial fitted to C1. If R1 and R2 are not equal the ratio between value of the unknown capacitor C2 and that of the known capacitor C1 is determined by the ratio R1/R2 in the same way as for a simple resistance bridge. By switching in various ratios for R1/R2, a wide range of capacitance values can be measured with this bridge.

One problem with this type of capacitance bridge is that it is difficult to obtain variable capacitors with values much above 500 picofarads. Even when using the ratio arms to increase the range covered it will be difficult to make measurements of capacitors much bigger than 100 nF with any degree of reliability. A further problem is that most readily available variable capacitors are designed for use as tuning capacitors in radio receivers. These capacitors are designed to give a linear frequency scale on a receiver and they have a nonlinear scale for capacitance against shaft rotation. The result is that the bridge will also have a nonlinear scale.

By re-arranging the capacitance bridge circuit as shown in Figure 3.13 we can remove the need for the variable capacitor. In this

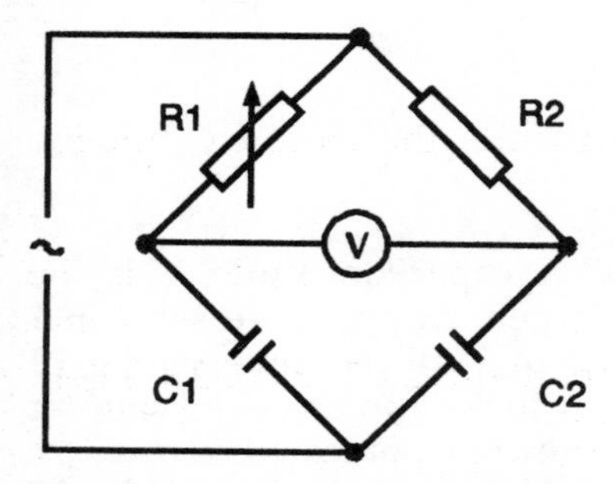

Fig 3.13 Modified capacitance bridge giving wider range.

scheme a fixed capacitor C1 is used as the reference against which C2 is compared and resistor R1 is made variable. In effect we have made the ratio term R1/R2 variable and this is used to balance the bridge. The value of C2 can now be read off from the scale of R1. If the maximum value of R1 is 10R2 the capacitance values up to 10 times the value of C1 could be measured.

To handle a wide range of capacitance values a series of different values of capacitor C1 can be switched into circuit. With this type of circuit it should be possible to measure capacitance values from a few picofarads up to several microfarads. One advantage is that a linear scale precision potentiometer can be used for R1 to give a linear capacitance scale.

Digital capacitance meter

Although the use of a bridge to find the value of a capacitor is quite effective a more convenient instrument for this purpose is the digital capacitance meter. The big advantage of this instrument is that, unlike a capacitance bridge, there is no balance adjustment. The capacitor to be measured is connected to the test terminals and its value is displayed as a digital number on the instrument's display.

The digital capacitance meter is really just a variation of the basic single ramp digital voltmeter. In the voltmeter the capacitor which is charged up has a known value and its voltage is compared with the input voltage. In the capacitance meter the capacitor under test is charged at a known constant current and its voltage is compared with a known reference voltage. The time the capacitor takes to charge to this voltage level is then measured and this will be proportional to the capacitance of the capacitor under test.

The time taken for the capacitor to charge to voltage V is

$$t = \frac{C \cdot V}{i}$$

where i is the charging current, C is capacitance and V is the voltage to which the capacitor is charged. Suppose we set V at 1 volts and i at 100μA. For a capacitor of 1 μF the time will be 10 milliseconds. If we now use a clock of 100kHz to drive the counter the resultant count would be 1000. If the capacitor tested were 0.5 μF it will charge up twice as fast and the new time period would be 5 ms giving a count on the display of 500. Thus assuming the maximum count were 1999 this meter would read capacitances up to 1.999 μF to the nearest nanofarad. By reducing the charging current to 10 μA the capacitance range would be from 0 to 199.9 nF measured to the nearest 100 pF.

A typical commercial digital capacitance meter provides ranges of 0 – 200 pF, 0 – 2nF, 0 – 20 nF, 0 – 200 nF, 0 – 2 μF, 0 – 20 μF and 0 – 200 μF. Some handheld digital multimeters also provide facilities for capacitance measurement and will usually have a similar set of capacitance ranges. In these instruments, the internal logic is switched when the mode is changed from voltmeter to capacitance meter. On most instruments it is important to ensure the the capacitor is discharged before connecting it to the meter to prevent possible damage to the meter.

Measuring inductances

In theory, inductances could be measured by using a variation of the simple capacitance bridge but with inductors substituted for capacitors. Unfortunately such a bridge is not a very practical proposition since it requires the use of an accurately calibrated variable inductor and such components are not readily available.

An alternative bridge arrangement for measuring the value of an unknown inductor is the Maxwell bridge shown in Figure 3.14.

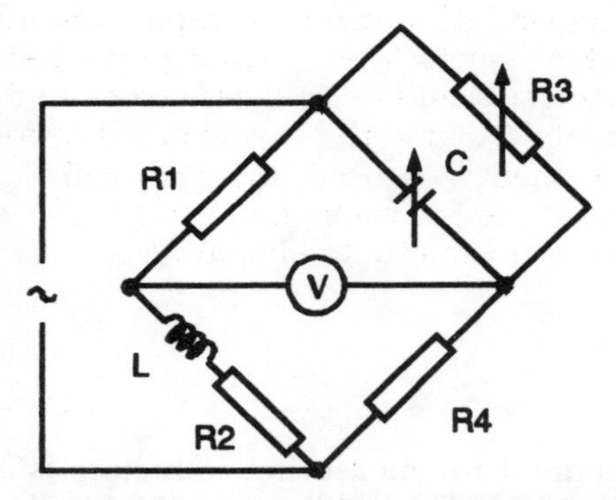

Fig 3.14 Maxwell bridge for inductance measurement.

Here the inductor arm is diagonally opposite a capacitor arm and when balance occurs the unknown inductance is given by

$$L = R1 \cdot R4 \cdot C$$

Normally the inductor will also have some series resistance R2 and for an accurate null reading this will also need to be balanced out. This can be done by adjusting the variable resistor R3 in parallel with the capacitor C. Initially R3 is set at maximum and the bridge is adjusted for a null by varying the value of C. When this rough null has been found R3 can be adjusted to balance the resistive component of the inductor. Further small adjustments of C and R3 may then be made to achieve the best null.

The series resistance of the inductor is then given by

$$R2 = \frac{R1 \cdot R4}{R3}$$

Although the Maxwell bridge can work reasonably well the value of the variable capacitor is limited to about 1000pF if a conventional air-spaced tuning capacitor is used. With an audio frequency signal source 1000pF has a rather high impedance. If the inductor to be measured has a low value then the ratio arms have to be set at a high ratio and this can lead to poor results. One solution is to increase the frequency at which the bridge is operated.

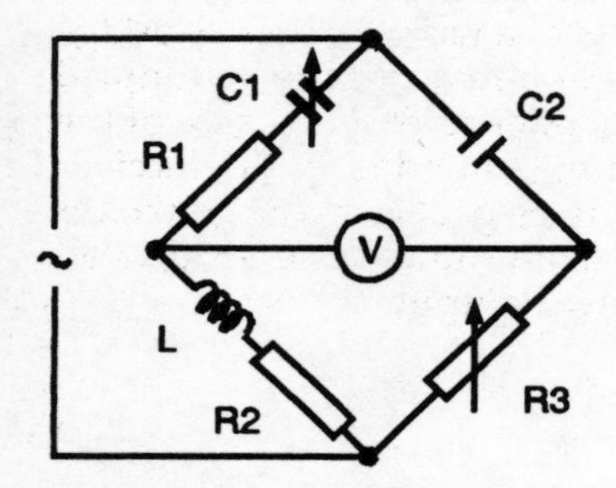

Fig 3.15 Owen bridge circuit.

An alternative bridge arrangement for measuring inductors is the Owen bridge shown in Figure 3.15. This bridge uses the small capacitor C1 to balance out the resistive component R2 of the inductor. The inductive component is now balanced by adjusting R3 and the capacitor C2 has a fixed value. The capacitor C2 can now be switched to a set of preset values to provide a number of ranges of measurement.

When the bridge is balanced the value of the inductor is given by

$$L = R1 \cdot R3 \cdot C2$$

which is basically the same result as for the Maxwell bridge.

The resistive component R2 is given by

$$R2 = \frac{R4 \cdot C2}{C1}$$

Radio frequency bridges

The basic audio frequency bridge circuits can also be used at radio frequencies for measurements of inductance, capacitance and impedance. For radio frequency operation the signal source can be a signal generator which will produce a sine wave signal at the desired radio frequency. A peak reading diode detector using a semiconductor diode may be used as the null detector. An alternative balance detector arrangement is to couple the bridge output to the antenna input terminals of a radio receiver which is tuned to the frequency of the signal generator. Some care is needed with coupling the signal to the receiver to avoid stray signal pickup.

For optimum results the components of a radio frequency bridge will need to be screened and the generator may need to be coupled to the bridge via a ferrite cored transformer so that one side of the receiver connection to the bridge can be grounded. If the receiver is designed for amplitude modulation the signal from the generator should be amplitude modulated with an audio tone so that an audible signal is available at the receiver output.

The Q meter

Accurate measurements of inductance and capacitance using bridges can become difficult when radio frequency signals are used to energise the bridge. This is due to the effects of stray capacitance and inductance in the bridge circuits. An alternative technique for making measurements at radio frequencies is to use resonance methods. One example of an instrument which uses this type of measurement method is the Q or magnification meter.

Let us examine a series resonant circuit consisting of an inductor L, a capacitor C and a resistor R in series with an AC voltage source of voltage E as shown in Figure 3.16. At most frequencies the

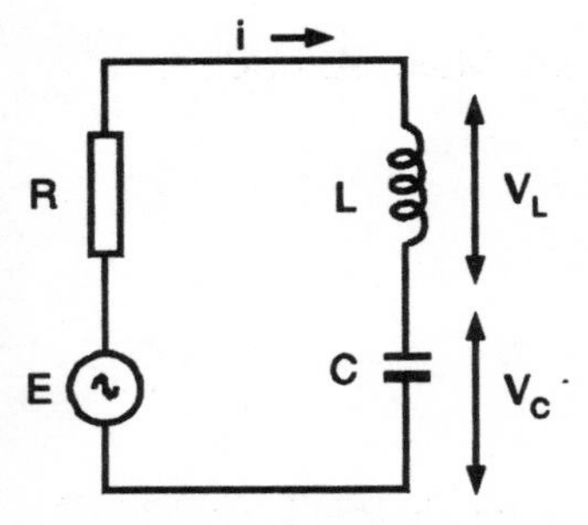

Fig 3.16 Series resonant LCR circuit.

inductive or capacitive reactance will be fairly high and this will limit the current flowing through the circuit to a low value. At one particular frequency however the inductive reactance will be exactly equal to the capacitive reactance and since these are of opposite sign they will cancel out leaving only the resistive component R. The frequency at which this effect occurs is the resonant frequency of the circuit. If the value of R is small then at resonance a large current will flow through the circuit. The current will cause a large voltage to be developed across the inductor L and an equal but opposite voltage to appear across capacitor C. Because the voltage across the inductor at resonance is much higher than the voltage injected into the circuit by the generator the inductor is said to have a magnification factor which is usually referred to as the Q of the inductor.

Since the current I flows through both R and L, the voltage across the resistor will be equal to I . R and the voltage across the inductor will be $I \cdot X_L$. At resonance the reactances cancel out so the generator voltage becomes the voltage across the resistor R. The magnification factor Q is the ratio of the voltage across L to that across R so it will be given by

$$Q = \frac{V_L}{E} = \frac{I \cdot X_L}{I \cdot R} = \frac{X_L}{R}$$

where V_L is the voltage across the inductor L at resonance, E is the voltage produced by the generator and X_L is the reactance of the inductor L at resonance.

The basic arrangement of a Q meter is shown in Figure 3.17. Here a voltage is generated across a small resistance r by passing a known value of current through the resistor. The signal used is a radio frequency sine wave and in many of the older versions of the Q meter the current flowing into r is measured by using a

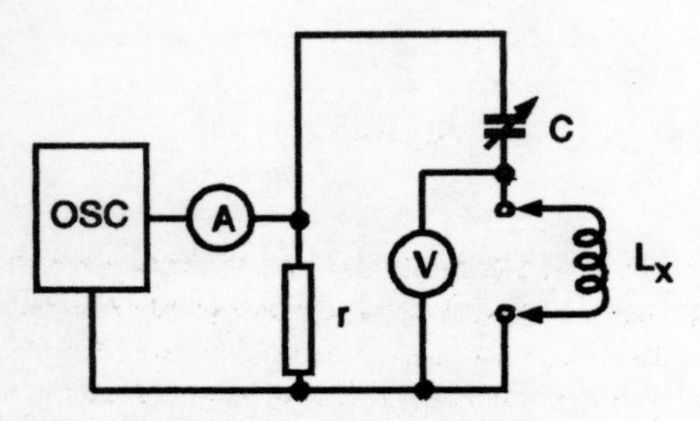

Fig 3.17 Basic arrangement of a Q meter.

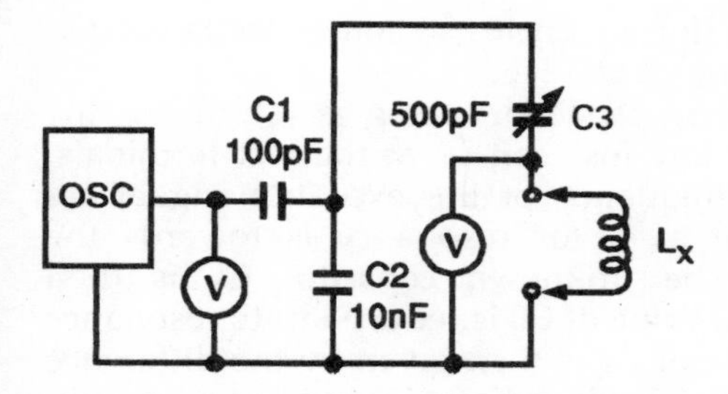

Fig 3.18 Capacitive divider scheme for signal injection.

thermocouple type milliammeter and is preset to produce a known voltage across r.

An alternative scheme for injecting a signal into the resonant circuit is to use a capacitive voltage divider scheme as shown in Figure 3.18. Here the oscillator signal is fed to capacitors C1 and C2 where the value of C2 might typically be 100 times the value of C1. The capacitors act as a potential divider for AC signals and the voltage across C2 will be approximately 1/100 of the voltage applied across the two capacitors. Since the capacitor C2 is also typically some 10 times the value of the tuning capacitor C3 it has little effect on the tuning of the circuit and if desired could be accounted for in the calibration of the scale on capacitor C3. With this type of circuit a voltmeter is connected across the oscillator input signal and this would be adjusted to some standard value such as perhaps 1V.

The coil to be measured L_x is connected to the test terminals and is then tuned to resonance by adjusting the variable capacitor C3. The voltage appearing across the capacitor is measured by an electronic voltmeter circuit. At resonance the voltage measured will rise to a maximum level. The voltage across the capacitor will have the same amplitude as that across the inductor. If we arrange that the voltage level injected into C1 is 1V then the voltage measured across the tuning capacitor will be the Q value of the coil multiplied by the ratio C1/C1+C2 and for the circuit shown this will be approximately Q/100. In practice there can be small errors due to the stray capacitances in the test circuit and self inductance of any test leads used. Self capacitance of the coil being measured can also affect the readings if it is comparable with the value of the capacitor C3 being used to resonate the circuit. Problems due to strays become increasingly important at higher radio frequencies and at VHF frequencies so some care is needed in using this type of instrument at VHF.

To measure an unknown capacitor using a Q meter the instrument is fitted with a low loss coil L_X at the test terminals. The value of L_X and the frequency of the excitation signal are chosen so that the setting of C for resonance is towards the maximum value of C3. The unknown capacitor C_X is then connected across C3 and the value of C3 is reduced until resonance is again achieved. The value of C_X is then given by the difference between the two settings of C3 when C_X is in and out of circuit. The capacitor C3 is normally calibrated directly in capacitance. Some Q meters have two capacitors connected in parallel to make up the variable capacitor C3. One capacitor has a large maximum value of perhaps 500 pF whilst the second is much smaller at perhaps 50 pF maximum. The advantage of this arrangement is that small values of unknown capacitance can be measured more accurately by using the smaller vernier capacitor for making the difference measurement. In this case the vernier capacitor would initially be set at its maximum value and resonance would be achieved by adjusting the main variable capacitor. The unknown capacitor is then added to the circuit and the vernier capacitor is adjusted to bring the circuit back to resonance. The value of the unknown capacitor is then given by the change in value of the vernier capacitor needed to bring the circuit back into resonance.

Measuring the value of inductance of a coil using a Q meter is a slightly more complex process. Since it not practical to produce an accurately calibrated variable inductor an indirect method of measuring the inductance of the unknown coil has to be used. The inductor being measured will in fact have some self capacitance and at some frequency f_0 the inductance will resonate with its own self capacitance. At this frequency the coil will simply appear to be a high value resistance.

First the coil L_X is connected to the Q meter and with C3 set near its maximum value the frequency of the signal which is driving the meter is adjusted to make the circuit resonate. The bridge frequency is then increased to about 8 times its current value and the unknown coil is replaced by a low loss inductor of a value which will allow the circuit to be resonated at the new operating frequency. The unknown coil is then reconnected so that it is in parallel with the tuning capacitor C3. The capacitor is then adjusted for resonance and a note is made of the direction of change needed. If the capacitor has to be increased the oscillator frequency should be increased slightly and then the circuit resonated again with L_X disconnected. If the capacitor value has to be reduced then the frequency should be reduced slightly. The unknown coil is again connected and further frequency adjustments made until the addition of the unknown coil L_X has no effect on the resonance of the meter circuit. At this new frequency of operation the unknown inductor L_X is resonating with its own self capacitance and therefore appears as a simple resistance as far as the Q meter is concerned so that there is no alteration of the resonance of the meter circuit.

If the original resonance frequency is f_1 and the new resonance frequency, where the coil L_x resonates with its own self capacitance, is f_0, then the self capacitance C_o of the inductor under test is given by

$$C_o = (f_1/f_0)^2 \cdot C$$

where C is the value of the meter tuning capacitor needed to resonate the coil at frequency f_1.

Having found the self capacitance of the inductor the value of inductance L_X is given by

$$L_x = \frac{C_o + C}{(6.28\ f_1)^2}$$

Another resonance method

If a Q meter is not available it is possible to measure the Q of a resonant circuit by using another resonance method. An oscillator is loosely coupled to the tuned circuit under test and the voltage developed across the tuned circuit is measured using an electronic voltmeter or a digital meter which has a high input impedance so that it does not load the tuned circuit. The basic circuit is shown in Figure 3.19.

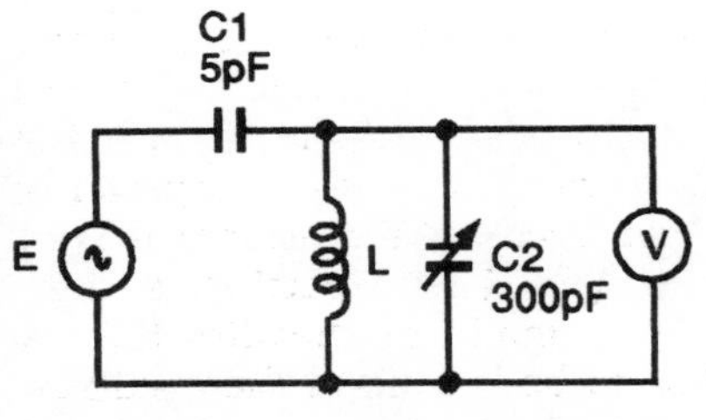

Fig 3.19 Loosely coupled resonant circuit.

The oscillator is set to the frequency at which the Q is to be measured and the tuned circuit is adjusted until it resonates. This will be indicated by a maximum reading on the voltmeter. The oscillator level may then be adjusted to give a reading of say 1V on the voltmeter. The frequency of the oscillator is then increased slightly until the voltmeter reading falls to 0.707V and the new frequency is noted. Next the oscillator frequency is decreased until it is below the resonance frequency and again the voltage has fallen to 0.707V. This new frequency is also noted.

The Q of the circuit is now given by

$$Q = \frac{f_r}{\delta f}$$

where f_r is the resonant frequency, where the output voltage is maximum and δf is the difference in frequency between the points where the output voltage falls to 0.707 of the value at resonance.

For best results the frequency settings should be measured using a digital frequency meter connected directly to the oscillator output. If the test is carried out at radio frequencies it would be possible to use an accurate radio receiver to measure the frequencies of the oscillator. If the receiver has a beat frequency oscillator for CW reception this should be turned on and the receiver is then tuned for zero beat with the oscillator signal.

4 The oscilloscope

So far we have looked at meters which give a picture of the static levels of voltage or current. For more complete tests on the operation of a circuit we need to be able to examine the way in which signals vary with time. This involves displaying a graph of the signal being examined against a base of time, and the instrument employed for this purpose is the oscilloscope.

The basic block diagram for a simple oscilloscope is shown in Figure 4.1. The main feature is a display screen on which a bright dot is made to travel horizontally across the screen at a fixed speed. The vertical position of the dot on the screen is moved in sympathy with the instantaneous level of the signal being examined. The dot therefore traces out a line which shows how the signal varies as

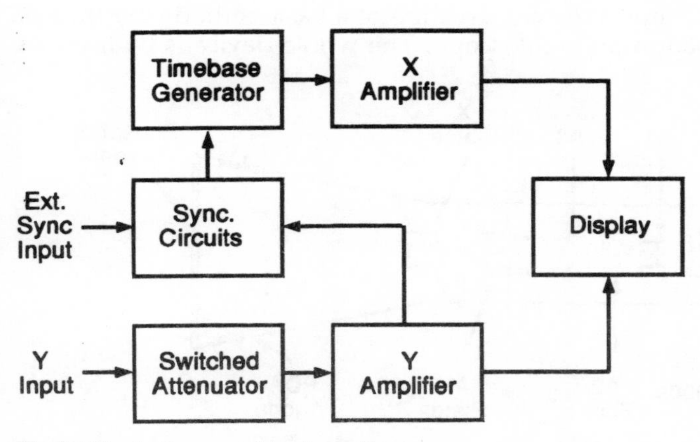

Fig 4.1 Block diagram of basic oscilloscope.

the dot moves across the screen. If the sweep across the screen is repeated at a rate of more than about 16 sweeps a second the image on the screen will appear to be a continuous line. This effect is caused by persistence of vision in the observer's eye. By adjusting the speed at which the dot moves we can arrange to pick out and display a pattern of one or more cycles of any repetitive signal such as a sine wave or a train of pulses.

The horizontal movement of the dot on the screen is controlled by a timebase generator whose output is amplified by the X amplifier before being used to drive the display unit. The input signal being examined is amplified by a second amplifier, called the Y amplifier, and is used to control up and down motion of the displayed spot.

Both X and Y amplifiers are fitted with gain controls which allow the size of the image on the screen to be adjusted. Part of the signal from the Y amplifier may also be injected into the timebase circuit to allow the horizontal motion of the spot to be synchronised to the waveform being examined so that a steady display is produced on the screen.

The cathode ray tube

In most oscilloscopes the display is produced by using a cathode ray tube which has some similarity to the picture tube in a television receiver. In recent years a few portable oscilloscopes have been produced which use a liquid crystal type display screen instead of a cathode ray tube.

Figure 4.2 shows the construction of a basic cathode ray tube of the type used in an oscilloscope. The whole device is built into a

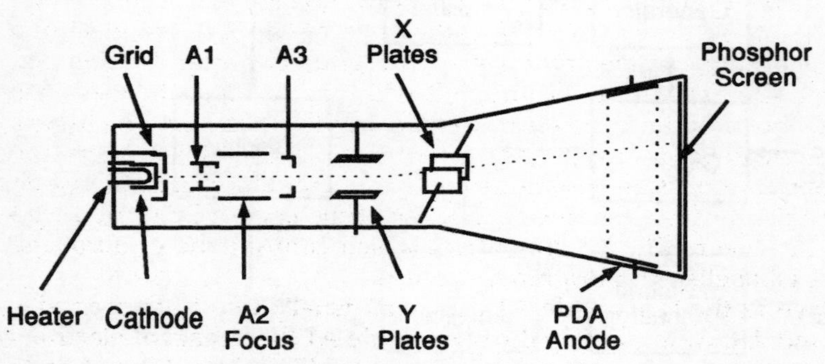

Fig 4.2 Basic construction of a cathode ray tube.

glass envelope which has all of the air pumped out and is then sealed to leave a vacuum inside the tube. One end of the cathode ray tube flares out and ends in a flat glass face plate which has its inside surface coated with a phosphor material which emits light when struck by a stream of electrons. This layer of phosphor forms the display screen on which waveforms can be viewed.

At the opposite end of the cathode ray tube is an electron gun which produces a stream of electrons. The electrons are focused by electrodes in the electron gun so that they all arrive at the same spot when they reach the phosphor screen coating. When the electrons strike the phosphor material it uses fluorescence to produce a small spot of light on the screen. Other electrodes in the electron gun are used to bend the path of the stream of electrons so that the spot produced on the screen can be moved to any part of the display area.

Most of the older types of cathode ray tube were built into a circular glass tube which flared out to a circular display screen. The typical screen size for these tubes was around 125mm in diameter but the working area of the screen was usually restricted to a rectangle 100mm wide by 80mm high. Modern cathode ray tubes for oscilloscopes generally have a round section for the neck of the tube, which contains the electron gun, but the flared part of the tube has a rectangular section and the faceplate is also rectangular. For these tubes the faceplate is slightly larger than the 100 x 80 mm working area of the screen. The main advantage of these rectangular face tubes is that they take up less panel space on the instrument and allow a more compact unit to be produced.

In the electron gun a stream of electrons is generated by heating a cathode element which is coated with barium oxide. The electrons emitted by the cathode are attracted towards a ring type anode by applying a fairly high positive voltage to the first anode A1. As they travel from the cathode to anode A1, the electrons pass through an electrode called the grid. If the grid is biassed negative relative to the cathode it will tend to repel the electrons back towards the cathode thus reducing the number which pass through to reach anode A1. Altering the grid voltage alters the density of the stream of electrons that eventually reach the screen and this in turn alters the brightness of the spot of light produced on the screen. The potentiometer which controls the grid voltage level is labelled the Brightness control.

Beyond the first anode A1 there are usually two further anodes A2 and A3. After passing through anode A1 the stream of electrons will tend to be diverging from the centre line of the tube. Anode

A2 is a cylindrical electrode designed to generate an electric field pattern which bends the diverging paths of the electrons in the beam and directs them all back towards the centre line of the tube. By adjusting the voltage of A2 relative to A1 and A3 the stream of electrons can be focused so that all of the electrons arrive at the same point when they reach the phosphor screen. In effect the electric fields produced between A2 and its adjacent anodes act in the same way as a lens in an optical system to produce a finely focused spot of light at the point where the electron beam hits the screen. The potentiometer which controls the A2 voltage level is labelled as the Focus control.

Beyond anode A3 there are two pairs of deflector plates which are used to move the position of the beam around the screen. The vertical or Y plates are mounted so that one plate is above the centre line of the beam and the other plate is below the beam. When a voltage is applied between the two Y plates the electrons in the beam are attracted towards the positive Y plate and repelled by the negative Y plate. The result is that the path of the electron beam is bent up or down in the direction of the positive Y plate. If the upper Y plate is made positive then the spot on the screen moves vertically towards the top of the screen. If the lower Y plate is positive, the spot moves down toward the bottom of the screen. The second pair of deflector plates are arranged horizontally with one plate at each side of the beam. These are the horizontal or X plates and are used to move the spot to the left or right on the screen.

The deflection sensitivity of a cathode ray tube is normally defined in volts/cm and represents the number of volts that need to be applied between a pair of deflection plates to produce a 1 centimetre movement of the spot on the screen. The deflection sensitivity depends upon the accelerating voltage applied to anode A3 of the gun. As the A3 voltage is increased the velocity of the electrons in the beam increases and a higher deflection voltage is needed to move the beam over the same distance. The figures quoted for deflection sensitivity apply only to a given value of voltage on A3.

To obtain a bright sharp trace on the screen the electron beam needs to be accelerated to a high velocity when it hits the phosphor. More brightness can be achieved by applying a higher voltage to the third anode of the gun but this results in lower deflection sensitivity on the X and Y plates. With older types of cathode ray tube the choice between deflection sensitivity and trace brightness was always a compromise, and oscilloscopes using these tubes tended to give poor display brightness when high speed waveforms were examined.

To overcome this brightness problem modern tubes are fitted with a post deflection accelerator (PDA) anode. This PDA electrode is a spiral graphite coating located on the inside of the glass envelope a short distance behind the screen. In this type of tube the A3 anode of the gun is operated at about +800V to 1000V to give high deflection plate sensitivity. The PDA electrode is operated at voltages from about +5000V up, perhaps +16kV, and this accelerates the electrons to produce a bright sharply focused trace on the screen. Since the main part of the acceleration of the beam is now carried out after it has been deflected by the X and Y plates the voltage levels needed for deflection can remain low but a bright trace is still produced on the screen.

The deflector plates normally operate at a voltage level which is the same as that of anode A3. For deflection one plate goes positive relative to A3 and the other negative relative to A3. If the cathode of the electron gun were operated at ground level the deflector plates would be at a voltage of about +1000V and this can present problems when coupling signals from the X and Y amplifiers to the plates. This problem is solved by setting the A3 voltage at about +100V and operating the cathode and grid at a level of −900V. The focus anode A2 will also be at quite a high negative potential. The Brightness and Focus control potentiometers are now at high potentials relative to ground but can be fitted with insulated shafts to prevent any hazard to the operator of the oscilloscope.

The Y amplifier

The voltage swing required at the Y deflection plates of the cathode ray tube is usually of the order 5V per centimetre on a modern cathode ray tube. For a signal trace using the full height of the screen this means that a signal of some 40V amplitude would be required. For typical applications we need to be able to display signals of perhaps a few hundred mV and expand them to fill the screen. To achieve this an amplifier is included between the Y signal input terminal and the Y deflector plates.

The basic arrangement of the Y amplifier is shown in Figure 4.3. The input signal is passed through a switched attenuator to a high input impedance pre-amplifier stage. The attenuator selects the overall sensitivity in terms of input voltage needed to produce a

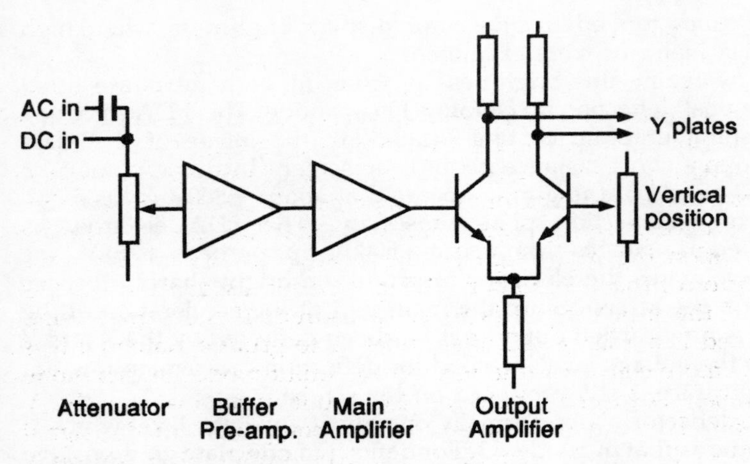

Fig 4.3 Block diagram of typical Y amplifier.

deflection of 1cm on the screen. The signal from the pre-amplifier then passes through a second amplifier stage which provides the voltage gain required when a high input sensitivity is selected. The amplified signal is then fed to a balanced output amplifier which drives the Y deflector plates on the cathode ray tube. On all modern oscilloscopes the Y amplifier is of the directly coupled type so that it can respond to DC signals. A variable DC bias signal is usually applied to one of the inputs of the final amplifier and this allows the position of the trace to be moved vertically on the screen. The potentiometer which provides this bias signal is usually labelled Y SHIFT or VERT. POSITION.

The input signal to be examined may range in amplitude from a few millivolts to tens of volts. The overall gain of the Y amplifier is usually designed to provide a deflection sensitivity of perhaps 10mV per division on the screen when the input is unattenuated. To cope with the wide range of input signals a switched attenuator is included between the input terminals on the instrument and the Y amplifier input. This usually has sensitivity ranges in the 1, 2, 5 sequence ranging from 10mV/div up to perhaps 50V/div and these are selected by a rotary switch. In addition to the switched attenuator some oscilloscopes also have a potentiometer as a variable sensitivity control which permits intermediate values of sensitivity to be selected so that the size of the displayed waveform can be adjusted to any desired level. If this variable control is used the Y calibration selected by the attenuator switch is no longer valid. One end of the variable control is marked as the calibrate position and the control should always be set to this position when measurements of the input signal amplitude are being made.

The Y amplifier input has an input mode switch which selects AC or DC inputs or Ground. In the Ground position the switch connects the Y amplifier input to the ground or 0 volt line and this mode is used to set up a zero datum position for the trace. In the DC position the input signal is connected directly to the input attenuator of the Y amplifier. This position is used when the DC level of the input signal is to be measured. If the signal being examined has a large DC offset relative to ground potential it is usually more convenient to use the AC input mode. In this mode a capacitor is inserted between the input terminals and the Y amplifier. This capacitor blocks the DC component of the signal but allows the waveform of the signal to be examined. An example is where the oscilloscope is used to examine the signals at the anode of a vacuum tube or to look at the voltage ripple on a high voltage dc line.

On a typical modern oscilloscope the Y amplifier has an input impedance of about 1 MΩ shunted by a capacitance of the order 50pF. The input connector is usually a BNC type coaxial socket.

Y amplifier bandwidth

The Y deflection plates of the cathode ray tube are connected across the output circuit of the Y amplifier and they present a small capacitance which shunts the amplifier load resistors. At low frequencies in the audio range the capacitance of the plates has no effect but as the frequency rises the capacitive reactance presented by the plate capacitance falls in value. At some frequency this capacitive reactance becomes equal to the load resistance of the amplifier stage. The load impedance seen by the amplifier is reduced due to the shunt capacitive reactance and the effective voltage gain of the amplifier is reduced.

Ideally the amplifier should provide constant gain at all frequencies from zero to infinity. In practice this is not possible and the amplifier will have a flat response up to a relatively high frequency but then the gain will fall off as the frequency increases. The bandwidth of the Y amplifier is defined as the frequency at which the amplifier gain falls by 3dB from the low frequency value. Typical amplifier bandwidths range from perhaps 5 – 10 MHz for a simple hobbyist oscilloscope up to 50 – 100 MHz for a professional or laboratory oscilloscope. A good compromise for the amateur enthusiast and the small workshop might be an oscilloscope with a bandwidth of around 15 – 20 MHz. Such an instrument can handle all audio frequency and most video signal requirements.

The timebase generator

In order to produce a regular trace across the screen a repetitive waveform must be applied to the X plates of the cathode ray tube. This waveform is usually a linear sawtooth similar to that shown in Figure 4.4. During the sweep time the voltage between the

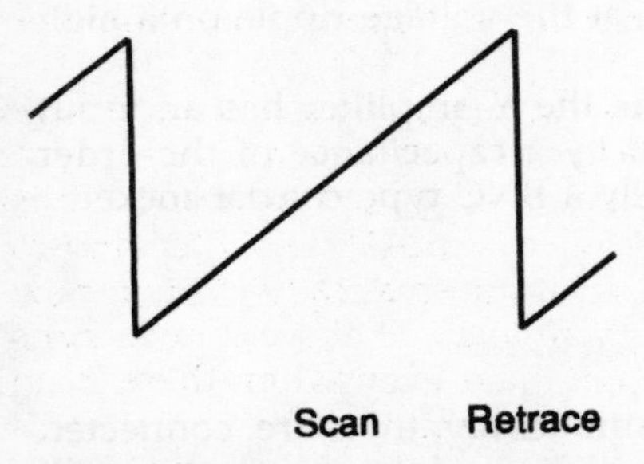

Fig 4.4 Sawtooth waveform produced by timebase generator.

plates is increased linearly with time to move the spot smoothly across the screen from left to right. At the end of this scan period the spot is returned rapidly to the left of the screen ready for the next scan.

The sawtooth waveform for driving the X plates of the cathode ray tube is produced by charging a capacitor at constant current to produce the linear sweep across the screen and then the capacitor is short-circuited to return the spot to the left hand end of the screen trace. The waveform is usually generated at low voltage and then applied to the input of the X amplifier which increases the voltage swing to a level suitable for driving the deflector plates. An X shift control applies a dc bias to one of the stages of the X amplifier and this allows the entire trace to be moved either to the left or right from its normal position. The bias control is normally labelled X SHIFT or HORIZ. POSITION.

The scan speed of the timebase is controlled by a rotary switch which selects a number of preset sweep rates. This switch is usually labelled TIME/DIVISION and the steps are usually organised in a 1, 2, 5 sequence. On a typical modern oscilloscope the sweep rates would be from 500ns/div to perhaps 2s/div. Thus on the 500ns/div range each division on the oscilloscope screen calibration would represent a time period of 500 nanoseconds. On most oscilloscopes the screen is marked off in 1cm squares but on some portable oscilloscopes the screen scale divisions may be smaller than 1cm so it is more convenient to have the calibration in time per division on the screen.

On some oscilloscopes it is possible to vary the scan speed on a selected switch range by using a potentiometer. This control has a position at one end which gives the correct calibrated speed for the current switch position. When taking measurements of time from the screen it is important to ensure that the speed potentiometer is set to the calibrate position.

The timebase generator is usually arranged so that it will produce a single sweep across the screen each time a trigger pulse is applied. In normal use the trigger pulse for the timebase is derived from the signal being examined. An AUTO switch allows the timebase circuit to be set so that it automatically starts a new scan as soon as the retrace is complete. The scan now runs repetitively so that a trace can be obtained even when there is no synchronisation signal applied to the circuit.

All modern oscilloscopes provide an X expansion facility. This is usually a switch which increases the gain of the X amplifier by a factor of either 5 or 10. Under these conditions the scan sweeps the dot off the screen over most of the timebase sweep period. By using the X SHIFT control it is possible to examine portions of the scan in closer detail.

Dual timebase systems

A problem that can occur in examining complex waveforms is that the area of interest is a small segment of the waveform near the middle of the horizontal sweep. If the timebase sweep speed is increased the result is that the portion of the wave to be examined will move off the right edge of the screen.

One solution to this problem is to increase the X amplification so that only a portion of the sweep is displayed on the screen. The X shift can then be used to move the expanded trace horizontally until the area of interest appears on the screen. Most oscilloscopes provide a x5 or x10 sweep magnification but this usually results in a lack of brightness in the displayed trace.

In examining the video signal for a television channel it may be desirable to be able to examine a particular scan line on the TV screen. This might be a requirement for examining the data lines for a teletext signal. This would involve magnifying the X sweep by perhaps 300 times and then selecting the appropriate line by using the shift control. The problem here is that the resultant display is rather dim and it can be difficult to judge exactly where the displayed segment is relative to the start of the sweep.

Most of the more expensive oscilloscopes provide a dual timebase system to allow examination of a part of the complete timebase scan. The main timebase is set up with a sweep speed which shows a complete cycle of the waveform. In the case of a video signal this might be a complete field scan. At a selected point along this sweep a second timebase generator is triggered which produces a much higher sweep speed and allows a segment of the trace to be displayed in more detail.

One scheme for using dual timebases allows selection of either the slow (A) timebase, the faster (B) timebase or a combination of both. Initially the A timebase is selected and the main display is set up and synchronised to give a steady picture. Next the A+B mode is selected. In this mode, the B timebase is triggered and generates a bright up pulse which increases the trace brightness for the part of the A scan where the B timebase is running. This brightened area of the trace can then be positioned over the part of the signal that is to be examined in detail by setting the B delay control. At the same time the length of the B scan can also be adjusted by setting the B timebase speed. Finally the B mode is selected. In this mode the trace on the screen is generated by the B timebase and it now fills the screen to allow detailed examination of a part of the original A trace waveform.

The synchronisation signals for the A and B timebases can be input separately. If the signal being examined were a television signal, the A timebase might be triggered by the TV field pulse and the B timebase be triggered by TV line pulses. In this case as the B delay time is adjusted the B timebase can be locked accurately to give a steady display of one particular line in the TV field scan.

An alternative mode of operation used on some oscilloscopes is that the scan speed is switched as the trace moves across the screen. Thus the basic trace is produced by the slower A timebase but when the trigger point for the B timebase is reached the drive to the X plates is switched to the B timebase and a section of the trace is expanded. At the end of this expanded portion of the trace the slower scan is switched back to the X plates. This has some advantages since the effect is like moving a magnifying lens over the main trace to expand a portion of it for examination.

Input probes

To avoid pick-up of stray signals on the oscilloscope test leads it is usual to employ screened cables for feeding the signals from the test point to the input of the Y amplifier. The capacitance between the core and screen of a typical 1 metre long input cable could be about 50pf which, when added to the 50pF input capacitance of the amplifier will give a total shunt capacitance of 100pF across the circuit being tested.

When the signals being examined have relatively low frequencies, such as the waveforms expected in an audio amplifier, the capacitance of the test lead usually poses no problem and has little effect on either the waveform of the signals being displayed or the circuit being tested. When high frequency signals or fast pulses are being examined, the capacitance between the core and screen of the input cable can affect the waveforms that are displayed and may upset the circuit being tested.

Suppose the circuit being examined is a video amplifier with a load impedance of 1000Ω and the signal being examined is a 10 MHz sine wave. If we assume that the shunt capacitance of the oscilloscope and its test lead is 100 pF then at 10 MHz the capacitive reactance which is shunted across the amplifier load resistor by the addition of the oscilloscope is about 300Ω. The effect of this capacitive reactance will be to reduce the actual signal level appearing at the amplifier output so that the waveform displayed is about one third of the amplitude than it would be with the scope disconnected.

If the signal applied to the video amplifier is a square wave at 10 MHz, the displayed waveform will become triangular in shape because the capacitor is unable to charge and discharge fast enough through the amplifier load resistor to be able to follow the 10 MHz square wave.

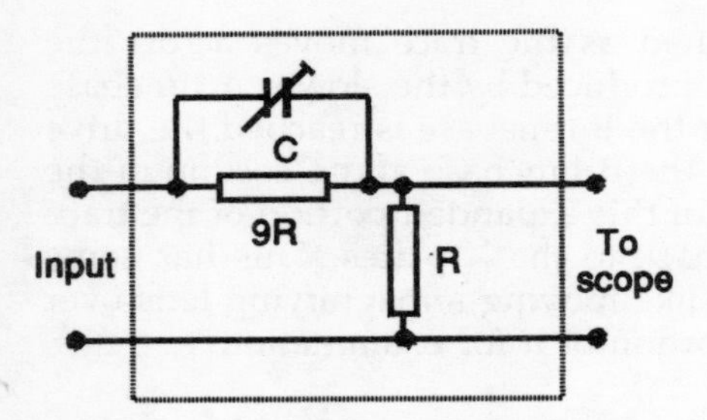

Fig 4.5 Circuit arrangement of simple input probe.

One way of overcoming this problem is to use a special probe at the input end of the test lead. This probe is usually arranged to act as a divide by ten attenuator and the circuit arrangement might be as shown in Figure 4.5. The dc component of the signal is attenuated by a pair of resistors forming a simple potential divider. To balance up the capacitive reactance a small series capacitor is connected across R1. The value of this capacitor is adjusted so that it has a capacitive value which is 1/9 that of the shunt capacitance of the lead and the oscilloscope amplifier input.

For example where the oscilloscope has a shunt capacitance of the order 50 pF, the series correction capacitor becomes approximately 5pF. Now when the probe is used to examine the video amplifier circuit it presents an effective reactance of around 3000Ω at 10 MHz and will therefore have much less effect on the signal being examined.

Probe tests

When a probe is included in the input line it is important to match the probe to the oscilloscope input. This is usually achieved by adjusting the small compensation capacitor in the probe to produce the correct results on a square wave input. Most oscilloscopes provide a square wave test signal for setting up input probes. This signal is applied to the probe input and the probe capacitor is then adjusted to give a correct square wave on the screen.

If the compensation capacitor in the probe is too large it will not produce the correct attenuation ratio for high frequency signals. In a square wave input this will give rise to overshoot on the edges of the square wave as shown in Figure 4.6(a). When the

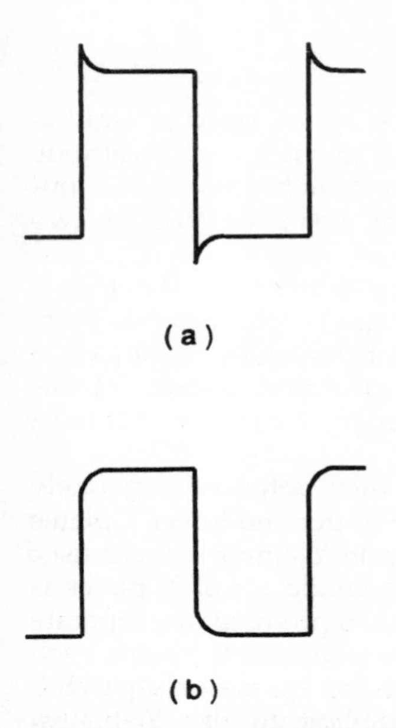

Fig 4.6 Effects of probe compensation adjustment.

compensation capacitor is too small the higher frequencies are attenuated too much and this produces rounded corners on the square wave as shown in Figure 4.6(b). With the correct setting of the compensation capacitor there should be no overshoot or rounding off on the edges of the square wave and the waveform is displayed correctly.

Multiple trace displays

If we have two signals present in a circuit and we wish to examine their relationship to one another in terms of their timing or phase, it would be very helpful if both signals could be displayed on the screen at the same time. This result can be achieved by using a special type of cathode ray tube which generates two separate electron beams each of which produces a trace on the screen. The alternative is to use a beam switching technique which produces two traces using a standard single gun cathode ray tube.

Dual beam tubes

The obvious scheme for displaying two input signals simultaneously on the screen of an oscilloscope is to use a special cathode ray tube which generates two separate electron beams. Each input signal is used to deflect its own electron beam to produce two separate traces on the screen.

In early Cossor dual beam scopes, such as the 1035, the special tube uses a single electron gun but the beam is split into two parts by a beam splitter plate mounted halfway between the Y deflection plates. This splitter plate is connected to the third anode. At this point in the tube the beam itself is not sharply focused so it is fairly easy to split it into two parts.

The two halves of the split beam are then deflected separately by the differences in voltage between the upper and lower Y plates and the central splitter plate. Two separate Y amplifiers are used to drive the two Y plates of the tube. A single set of X plates is used to deflect both beams simultaneously to produce two separate synchronised traces on the screen. The two parts of the beam each focus to a small spot on the screen but the spots are separated vertically according to the voltages applied to the Y plates. Although this scheme worked quite well, problems could occur due to interaction between the two parts of the electron beam.

Modern dual beam cathode ray tubes use two complete electron guns mounted side by side in the neck of the tube. Each gun includes its own set of deflector plates and produces its own separate trace on the screen. Separate brightness and focus controls are provided for each gun so that the two traces can be adjusted independently. In most oscilloscopes the two sets of X plates are tied in parallel and driven by a single timebase with a common X shift control which moves both traces simultaneously.

The dual gun technique is perhaps the most ideal in terms of flexibility since the two traces are completely independent and the system does not usually suffer from interaction problems between the two beams. The main disadvantage is that the dual gun tube is much more expensive to produce than a single gun type. It is possible to have more than two guns and cathode ray tubes have been produced with as many as eight separate guns to provide multi-trace displays for specialised purposes.

Two separate Y amplifiers are used to drive the Y plates in the two guns. These are usually labelled as CHANNEL 1 and CHANNEL 2, and each channel has its own shift and gain controls so that the two traces can have their position on the screen and signal amplitudes set independently.

Although the dual beam technique is the best for multi-trace displays, most modern oscilloscopes tend to use the less expensive single gun tube with a beam switching technique to produce a dual trace facility. There are in fact two approaches which may be adopted which are known as the chopper and alternate scan techniques. We shall first examine the chopper type beam switch scheme.

Chopper beam switch

Suppose we have a pair of signals at 50 Hz where one is a sine wave and the other a triangular waveform which are of the same frequency but not in phase with one another. These two signals can be fed to a switch circuit which selects the two signals alternately and feeds one of them on to the Y input of the oscilloscope as shown in Figure 4.7. Let us suppose that the switching rate is around 1000 times per second. If one cycle of the input signal is displayed across the screen then each signal is displayed as a set of 1 millisecond wide segments each separated by a blank space of 1 millisecond duration whilst the other signal is being displayed. The result is a display similar to that shown in

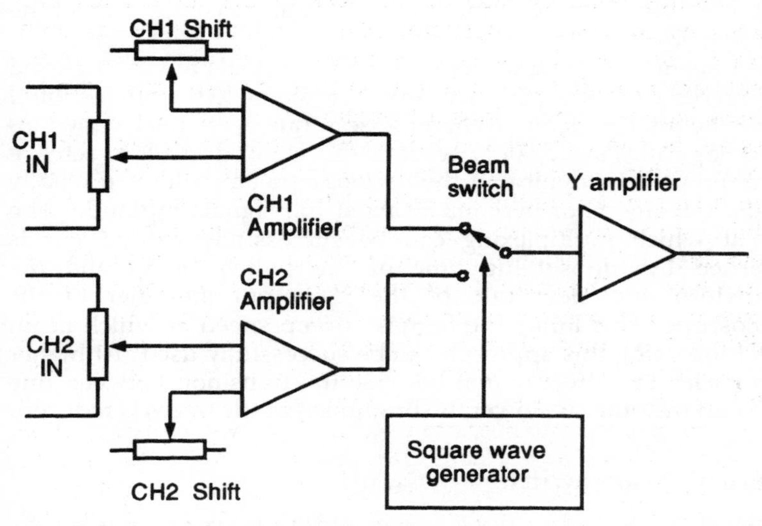

Fig 4.7 Block diagram of chopper dual trace scheme.

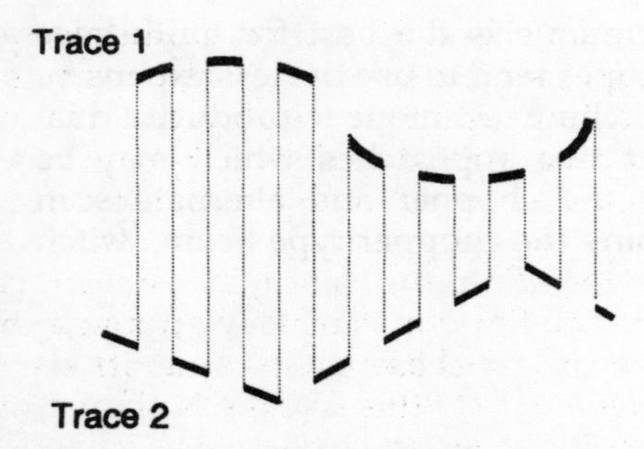

Fig 4.8 Dual trace display produced by chopper.

Figure 4.8. In a practical beam switching system the actual switching rate would be arranged to be of the order 100 or more times the frequency being displayed so that the trace becomes a series of closely spaced dots to give a good representation of the actual input waveform.

In most oscilloscopes the early stages of the Y amplifiers for the two channels are separate and the beam switching is applied in the final amplifier stages which drive the deflector plates. Each channel amplifier has its own attenuator system for gain control and a DC bias is applied to provide a Y shift for each channel. Since both signals can have a shift bias added the two traces on the screen can be moved independently to allow comparison of the two signals.

In a beam switching system where the signals are multiplexed during the scan the switch over from one signal to the other should ideally be instantaneous. When the scan rate is relatively slow, up to perhaps 1000 sweeps per second, this is not difficult to achieve since the switch circuit and the oscilloscope amplifier can readily handle transitions of perhaps 500ns at full signal amplitude. The rate at which multiplexing can be successfully carried out is governed by the settling time of the switch circuit and the bandwidth and slew rate of the Y output amplifier of the oscilloscope. This limits the highest sweep speed at which beam switching using this approach can be successfully used. At higher frequencies the effect is that because the transition between one trace and the other is slower the brightness of the traces is reduced.

Alternate beam switch

A different approach to beam switching which is more suitable for dealing with higher frequency signals is to multiplex the two input signals on alternate scans of the timebase. Thus on the first sweep across the screen the CHANNEL 1 signal is displayed and then on the next sweep the CHANNEL 2 signal is displayed. In this arrangement the multiplex switch operates during the retrace period of the timebase so that the speed requirements for the switch and the Y amplifier are no longer critical.

To achieve this mode of operation the multiplex switch is controlled by a flip-flop circuit. The flip-flop is usually arranged so that it changes state each time the timebase circuit ends its scan and starts a retrace. The arrangement for this scheme is shown in Figure 4.9 where a flip-flop circuit connected to operate as a divide by two circuit is used to drive the multiplex switch.

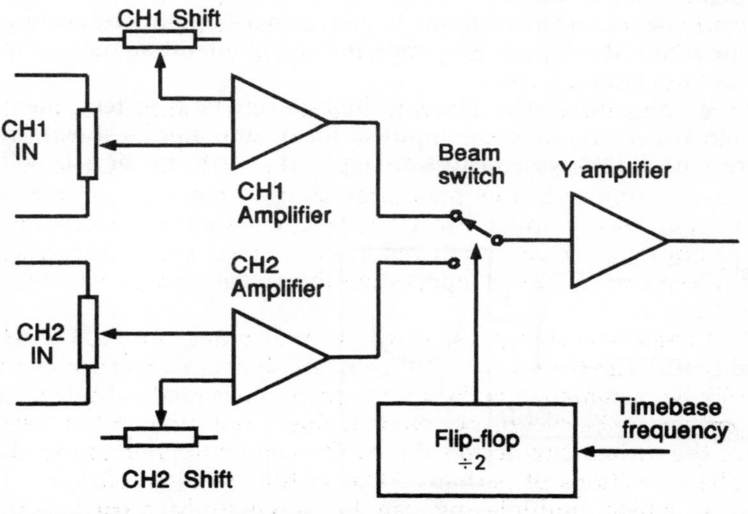

Fig 4.9 Block diagram of alternate trace system.

One useful advantage of the alternate sweep technique is that it becomes possible to examine two waveforms of different frequency and still display them as stable waveforms on the screen at the same time. This is achieved by ensuring that the timebase trigger signal is always derived from the signal that is to be displayed on each particular sweep. Thus on odd numbered sweeps the signal from CHANNEL 1 might be displayed and the sync. pulse is derived from CHANNEL 1 whilst on even numbered sweeps the signal from CHANNEL 2 is displayed and the sync. pulse is derived from the signal on CHANNEL 2. This facility is provided on some oscilloscopes but other simpler types synchronise both sweeps from the same signal source.

The alternate sweep scheme works well for repetitive waveforms but is not so good for non repetitive signals. Thus an event that occurred during the trace that displays the CHANNEL 1 signal does not occur simultaneously with an event displayed on the CHANNEL 2 trace and this can give rise to anomalous displays.

Algebraic displays

If we have a standard single trace oscilloscope and wish to display two signals simultaneously on the screen, one simple technique that could be used is to combine the two signals together by adding or subtracting them and then applying the resultant signal to the oscilloscope input.

The simplest way of combining the two signals is to feed them through two resistors to the input of the Y amplifier as shown in Figure 4.10. If the two resistors are equal then the two signals will

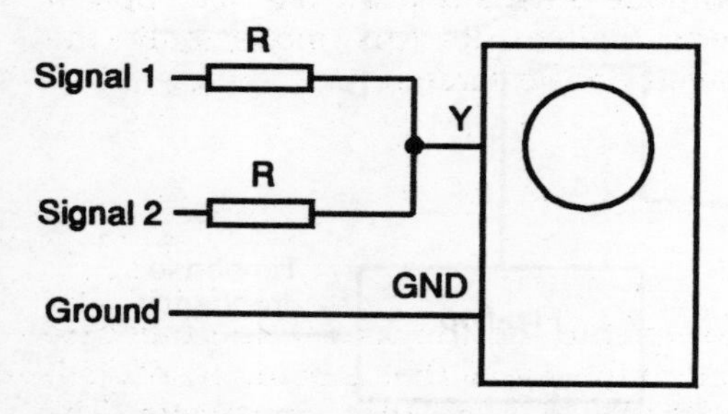

Fig 4.10 Signal addition using a resistor network.

add together but because of the potential divider action of the resistors, the actual signal at the oscilloscope input is half the amplitude of the combined input signals.

If the two signals are different in waveform as for example a sine wave and a pulse then it is possible to see and measure the timing relationship between them.

One problem with the simple resistor adding network is that it will only work successfully when the two signals are from low impedance sources. An alternative and better scheme is to use a

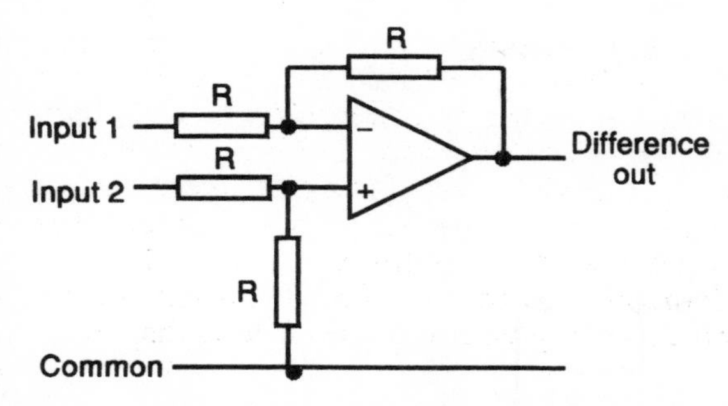

Fig 4.11 Signal subtraction using differential amplifier.

balanced input operational amplifier and to feed one signal into the amplifier's non-inverting input and the other signal to the inverting input as shown in Figure 4.11. In this case the output from the amplifier will be the algebraic difference between the two input signals. To obtain the algebraic sum, one of the signals is inverted before being applied to the amplifier input.

Many modern dual trace oscilloscopes provide the option of displaying algebraic sum or difference of the inputs on CHANNEL 1 and CHANNEL 2 and this mode is usually provided as an option on the CHANNEL 1 mode switch. In this mode only the CHANNEL 1 trace is displayed on the screen.

Trace synchronisation

In order to produce a steady picture on the screen, the timebase sweep must be synchronised to the signal that is being displayed. Suppose the signal being displayed is a 50Hz sine wave. The timebase sweep speed might conveniently be set at approximately 25 sweeps per second. If the timebase is continuously variable, its speed can be adjusted carefully until two cycles of the waveform appear on the screen. The problem is that the signal will not remain exactly in phase with the timebase sweep and the effect is that the displayed waveform on the screen will slowly drift across the screen. It will be difficult to examine the waveform properly when it is drifting in this fashion so we need a method for holding the display steady on the screen and this is achieved by a synchronisation circuit.

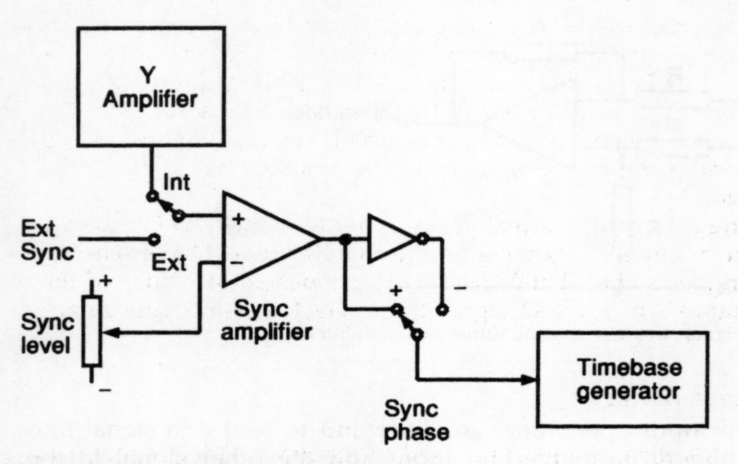

Fig 4.12 Block diagram of trigger circuits.

To give a steady picture the signal being displayed is used to provide a starting trigger for the timebase generator. The usual arrangement for the trigger circuit is shown in block form in Figure 4.12. A part of the signal from the Y amplifier is fed to one of the inputs of a high gain balanced amplifier which is usually referred to as the trigger amplifier. The second input is fed to a variable DC voltage level which can be set by a TRIG LEVEL control on the front panel. When the input signal rises above the preset trigger level it causes the output of the amplifier to swing rapidly from negative to positive. Similarly when the input signal falls below the preset level, the amplifier output swings from positive to negative. The output of the balanced amplifier becomes a square wave whose switching points occur where the input signal passes through the preset level.

One of the edges of the square wave from the trigger amplifier is used to trigger the timebase circuit so that it starts a new sweep across the screen. The start of the sweep is locked in time to a particular point on the waveform being displayed so that a steady picture is produced on the screen.

By adjusting the TRIG LEVEL control the trigger point can be moved relative to the displayed waveform. The trigger circuit can also be switched so that it responds to either a rising or falling part of the waveform. The oscilloscope will usually have a TRIG MODE switch which can select either + or − signal polarity for the trigger level. Thus for a sine wave if the TRIG MODE is set at + the trigger circuit responds to positive going signals and the sweep will start on the rising edge of the positive half cycle of the sine wave. As the TRIG LEVEL is increased the start will move futher up the rising edge of the sine wave. When TRIG MODE is set at − the trigger circuit responds to negative going signals and the sweep will now start on the falling edge of the negative half cycle of the sine wave.

The trigger circuit normally has a mode switch which allows either the Y amplifier signal or an external EXT signal to be selected as the trigger signal. Dual trace oscilloscopes allow either of the input channels or the EXT signal to be selected as a trigger input.

TV trigger mode

When an oscilloscope is to be used to display a video signal it is convenient to use the built in synchronisation pulses of the signal itself to trigger the oscilloscope timebase sweep. A television signal contains a series of built in pulses which are used by the television receiver to synchronise the picture scan to the received signal so that a stable picture is displayed. A line pulse occurs at the end of each horizontal scan line of the picture and a field pulse each time one complete vertical field scan is made through the television picture. The line and field pulses extend below the black level of the video signal as shown in Figure 4.13. In a television signal the field pulse is broken up into half line period segments as shown in Figure 4.14 because a single field consists of 312½ lines (262½ lines on the NTSC system used in North America). This scheme is used so that the scan lines on alternate fields are shifted by half a line so that they interleave with one another to produce the complete picture. Figure 4.14 has been simplified and in an actual signal, additional equalising pulses are included before and after the field pulse. Personal computers and video games often use a simpler scheme in which the field pulse is simply a long pulse extending over about 4 line periods.

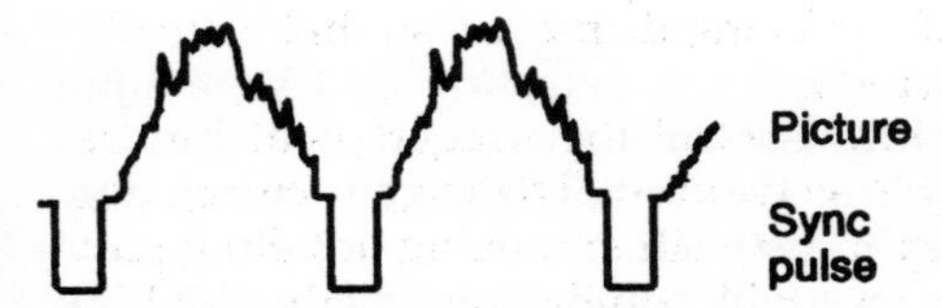

Fig 4.13 Typical video signal waveform.

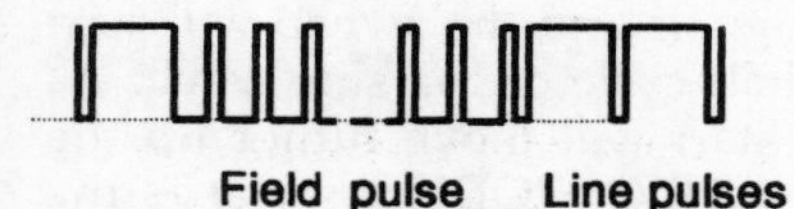

Fig 4.14 Field sync. pulses in a video signal.

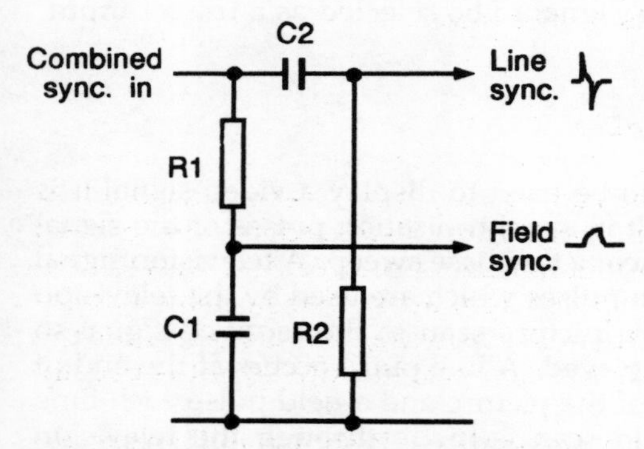

Fig 4.15 Circuit for separating line and field pulses.

The line and field pulses can be separated from the picture information by using a simple level sensing circuit since the synchronisation pulses are always below the black level of the video signal. Once the train of pulses has been separated the field and line sync pulses can be separated from one another by using an integrator and differentiator circuit as shown in Figure 4.15. The integrator acts as a low pass filter which allows the field pulse through but attenuates the line pulses. The differentiator produces a positive or negative spike each time the input signal switches levels, so it picks out the line pulses and the edges of the field pulses.

Most oscilloscopes have facilities for detecting and separating the TV line and field synchronisation signals and use these to trigger the oscilloscope timebase. This option is usually selected by a TV position on the TRIG MODE switch. Some oscilloscopes have separate switch positions marked TVL and TVF which allow the timebase to be locked to line or field pulses respectively.

Z input

Most oscilloscopes have a so called Z input which applies a signal to the grid or cathode of the display tube. This signal changes the grid bias on the tube and the brightness of the trace then changes in sympathy with the Z input signal.

The Z input is convenient for generating markers on the trace by applying short bright up pulses to this input which will produce bright spots on the trace at specific time points. An alternative approach would be to apply inverted pulses to the Z input to produce dark marker spots on the trace.

Amplitude measurement

The graticule in front of the cathode ray tube screen is used when making measurements on waveforms displayed on the oscilloscope. This consists of a grid of horizontal and vertical lines which, on most instruments, are spaced 1cm apart. For portable oscilloscopes with screens less than 10 cm wide, the divisions on the graticule may be smaller than 1 cm apart but the layout is usually the same.

The typical arrangement of the graticule is shown in Figure 4.16. The working area is usually 10 cm wide and 8 cm high and this area is divided into 1 cm squares. The central line of the vertical lines is marked off in 0.2 cm divisions to permit more precise

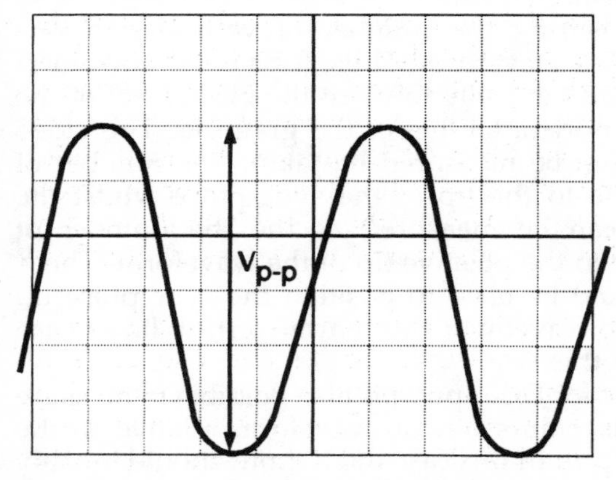

Fig 4.16 Measurement of peak – peak amplitude.

measurements of signal amplitude. Of the horizontal lines the one across the centre of the screen is marked off in 0.2 cm divisions for precise measurement of time.

Suppose we wish to measure the peak to peak amplitude of a sine wave. The range switch of the Y amplifier should first be set so that the amplitude of the wave is as large as possible whilst still remaining within the 10 x 8 graticule grid. The timebase speed should be adjusted to give at least two complete cycles of the waveform across the screen and the trigger level set to give a steady display.

To make the measurement, the Y POSITION control is adjusted to place the negative peak of the wave on the bottom horizontal line of the graticule. The X POSITION control is then adjusted to move the positive peak of the wave over the central vertical line of the graticule.

The amplitude can be determined by counting the number of divisions up the central vertical line starting from the bottom line of the graticule and going up to the point where the peak of the wave crosses the central vertical line. The result will be a number of whole screen divisions plus a fraction of a division at the top of the wave. To calculate the voltage, the result is multiplied by the V/division sensitivity of the Y amplifier range that is currently selected.

Figure 4.16 shows an example of amplitude measurement. Here the top of the positive half cycle is 6 divisions up from the bottom of the graticule and the Y input range selected is 5V/division. The peak to peak amplitude of the wave is 6 x 5 or 30V.

Other amplitude measurements on a waveform can be made in the same way by measuring the number of vertical divisions between two points on the waveform. Thus, for a video waveform the trace could be moved vertically to place the black level of the signal over the central horizontal line of the graticule. The video sync pulse amplitude can be measured by taking the number of divisions below this line to the tip of the sync pulse whilst the luminance amplitude can be measured as the the number of divisions above the line to the positive tip of the waveform. The X POSITION control should be used to position the sync pulse tip or the peak of the video wave over the central vertical line to see where the trace crosses this line.

If the Y amplifier is set for DC input it is also possible to measure the DC level of various points on the waveform relative to the common line of the Y input. In this case the Y input should initially be set to ground which can usually be done by setting a switch on the front panel. This effectively shorts out the input to the Y amplifier to give a zero datum line on the screen. It may be necessary to set the TRIG MODE to AUTO in order to get a trace under these conditions. The Y POSITION is then used to place the trace on the bottom line of the graticule. If the signal is known to be negative to ground then the trace could be set at the top line of the graticule. The Y amplifier is then switched back to DC input mode and the position of the signal relative to ground can be measured. Again the point on the signal to be measured should be placed over the central graticule line to allow a precise reading to be taken.

Amplitude calibration

Virtually all oscilloscopes provide a signal which can be used to check the calibration of the Y amplifier amplitude. In most cases this is a square wave derived from the supply mains and set to give an amplitude of 0.5V peak to peak. The signal is generally brought out on the front panel of the instrument.

To check the calibration of the Y1 input of the oscilloscope, connect its input lead to the calibrator voltage output socket. The timebase should then be set to about 20ms per cm and the Y voltage sensitivity is set for 100mV per div. This should give a signal of 8 divisions peak to peak deflection on the screen. If the amplitude is approximately correct then the Y amplifier is operating correctly. If there is a significant error then the preset gain controls inside the instrument will need adjustment to produce the correct Y sensitivity.

Current measurement

Normally the oscilloscope is used to display the voltage waveform applied to the Y input. For some tests however it may be desirable to display the current flowing in a circuit. An example of such an application would be to show the relationship between the voltage applied across a circuit and the current flowing through it.

The simplest way of displaying current on an oscilloscope is to break into the circuit and insert a small value resistor in series. The oscilloscope is then used to measure the voltage drop across the resistor which will be directly proportional to the current flowing through the circuit. This is shown in Figure 4.17 where the current

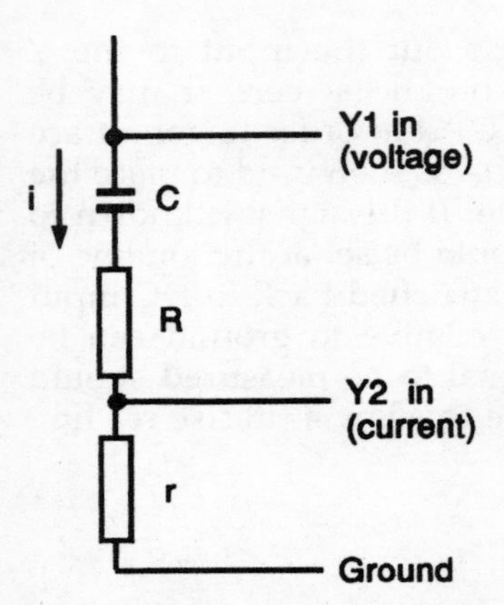

Fig 4.17 Current measurement using an oscilloscope.

in a series RC network is being measured. To avoid too much disturbance to the circuit being tested the series resistor used to monitor the current should have a value much lower than the impedance of the circuit.

Since the voltage developed across the resistor is small the Y amplifier used to monitor current will need to be set for a high gain. If a dual trace oscilloscope is being used the voltage across the circuit can be monitored using the second trace and the phase relationship between the current and voltage can then be examined.

An alternative technique for measuring current is to insert a transformer in circuit as shown in Figure 4.18. In this case the primary winding in series with the circuit would have a small number of turns and the secondary winding feeding the oscilloscope might have a larger number of turns to provide a step up ratio for the voltage. A resistor connected across the output winding is used to load the transformer so that it presents a low series impedance to the circuit being tested and the voltage output is then taken to the Y input of the oscilloscope to produce a display of the current flowing through the test circuit. The current in the circuit is then given by

$$I = \frac{N \cdot V}{R}$$

where V is the voltage measured on the oscilloscope, R is the resistance across the transformer secondary and N is the turns

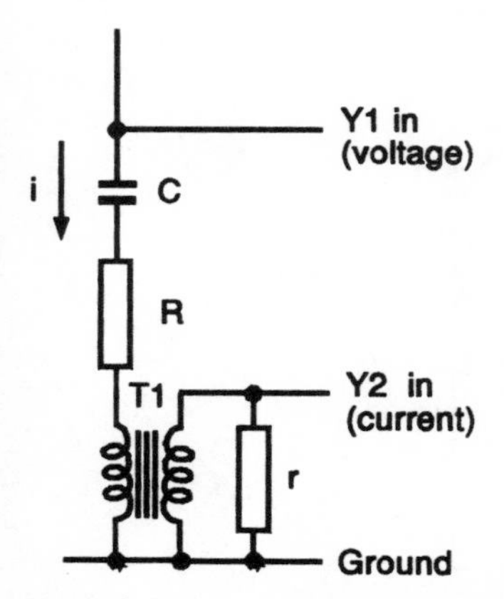

Fig 4.18 Current measurement using a transformer.

ratio between the secondary and the primary windings of the transformer.

Time measurement

The basic technique for measuring off periods of time between points on a waveform follows a similar procedure to that used for measuring amplitude.

Suppose we wish to measure the period of a sine wave. The timebase should be set so that one complete cycle of the waveform occupies as much of the horizontal space across the graticule area as possible. The Y sensitivity should also be set to give as large an amplitude as possible without going off the screen.

The trace is then shifted horizontally so that it starts about a quarter of a division to the left of the left hand vertical line of the graticule. The Y POSITION is then adjusted so that the point where the trace crosses the left hand vertical line is located on the middle horizontal line of the graticule. The number of horizontal graticule divisions is then measured to the point where the trace again crosses the central horizontal line at the end of the wave cycle. This is shown in Figure 4.19. The time period for one cycle of the displayed wave is given by the number of graticule divisions multiplied by the speed range of the timebase in ms/div or μs/div.

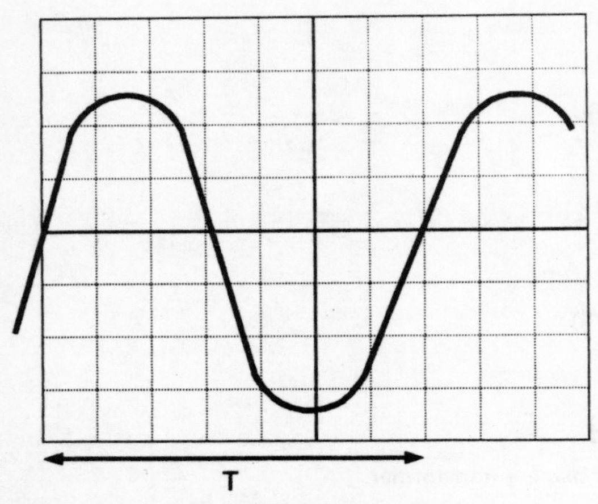

Fig 4.19 Measurement of a time interval.

When measuring high frequency signals where the timebase has been set to maximum speed there may be many cycles displayed across the screen. To obtain a better reading for the period of the signal it may be useful to switch on the X expansion which might well give a five or ten times increase in the width of each cycle. A point to watch is that if the X expansion is switched on, the time factor must be adjusted accordingly.

X calibration

The calibration signal used for checking the sensitivity of the Y amplifier is usually derived from the supply mains and can also be used for checking the timebase calibration. The calibration signal is applied to the Y1 input and the timebase is set for 2ms/div. The result should be a single cycle of the calibration signal spread across the screen. A further check with the timebase set at 5ms/div should produce a signal with 2½ cycles displayed and each cycle should occupy four divisions of the horizontal screen scale. If this is approximately correct, the timebase scale is set correctly. If there is a significant error, the timing presets inside the oscilloscope may need to be adjusted.

If an accurate signal generator is available which covers the range from about 100Hz to 10kHz this could be set up at frequencies of say 1kHz, 5kHz and 10kHz to check the timing accuracy of the higher speed ranges of the timebase generator.

Phase measurement

If we have two signals of the same frequency and wish to measure the phase difference between them this can be done by using a dual trace oscilloscope. One signal is fed to the CHANNEL 1 input and the other to the CHANNEL 2 input. The scan is synchronised to the CHANNEL 1 signal and the scan speed adjusted until there is about one cycle of the waveform displayed across the screen.

The CH1 POSITION is adjusted to place the CH1 trace so that it is centred about the horizontal axis of the screen graticule. The CH2 trace is then moved to place it over the CH1 trace. The X POSITION control is then adjusted to move the point where the CH1 trace crosses the horizontal axis to line up with the left hand vertical graticule line. The distance between the crossing point of the CH1 trace and the corresponding point on the CH2 trace is then measured along the horizontal axis of the graticule as shown in Figure 4.20. The total period for one cycle of the CH1 waveform should also be measured.

The phase shift will be the difference in position between the two traces divided by the total wave period and the result is multiplied by 360 to give phase in degrees.

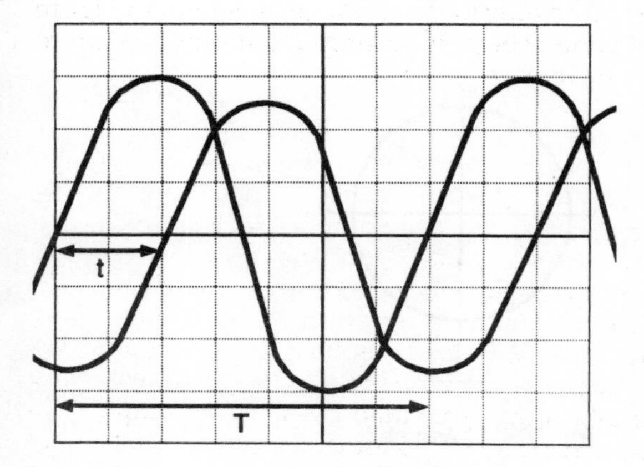

Fig 4.20 Measurement of phase difference.

Lissajous figures

One interesting way in which the oscilloscope can be used to compare the phase relationship between two AC signals is to apply one signal to the X plates of the tube and the other signal to the Y plates of the tube. This produces a display which is generally referred to as a Lissajous figure. On dual trace oscilloscopes there is usually a position of the TIME/DIV switch which selects the CH2 signal and applies it to the X amplifier. With this mode selected one signal is applied to the CH1 input and the other to the CH2 input.

When the two signals applied have the same frequency and are exactly in phase, the result will be a diagonal line on the tube which will run from the bottom left of the screen to the top right as shown in Figure 4.21(a). If one of the signals is now reversed

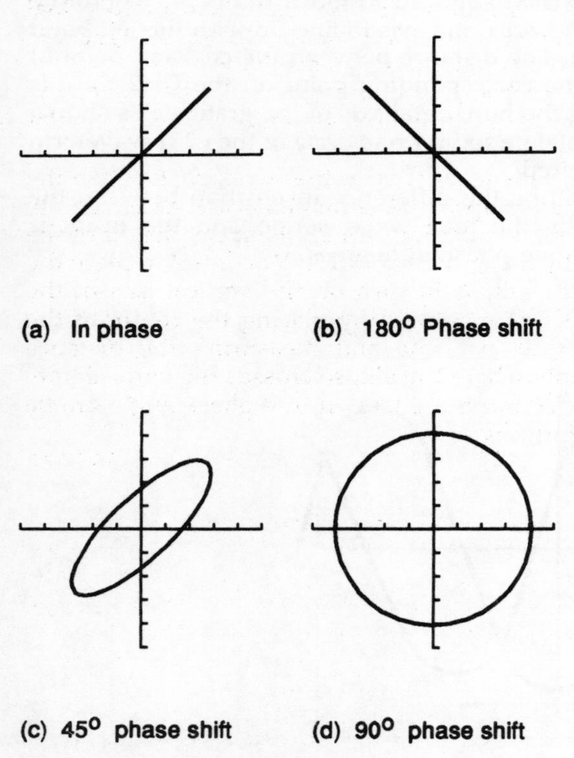

Fig 4.21 Typical Lissajous figure displays.

in polarity so that it is 180 degrees out of phase with the other signal, the result is still a straight diagonal line but now it will run from the top left to the bottom right of the screen as shown in Figure 4.21(b).

When the two signals are not quite in phase with one another the diagonal line changes to an ellipse running diagonally from bottom left to top right of the screen as shown in Figure 4.21(c). As the phase difference is increased the thickness of the ellipse will increase until it becomes a circle when the signals are 90 degrees out of phase as shown in Figure 4.21(d).

The above results assume that the signals being compared are sine waves which are of the same amplitude. It is also assumed that the deflection sensitivities of the X and Y circuits of the oscilloscope are the same. If the signal amplitudes or deflection sensitivities are not identical on X and Y then the resultant image will be stretched in the direction with the higher sensitivity. Dual trace oscilloscopes which use the CH2 channel for the X signal are usually matched to give equal X and Y sensitivities.

When the waveforms being examined are not sine waves the Lissajous display becomes distorted but generally follows a similar type of pattern.

To calculate the phase angle between the two signals when the Lissajous display is an ellipse, measurements are taken of the dimensions A and B shown on Figure 4.22. First the height of the ellipse can be found by using the X POSITION control to place the top and bottom of the ellipse in turn on the vertical axis of the graticule. The height B is measured by placing the centre of the ellipse over the vertical axis line and measuring the distance between the two points where the ellipse crosses the vertical line.

Once the values of A and B are known, the phase angle can be calculated from the formula

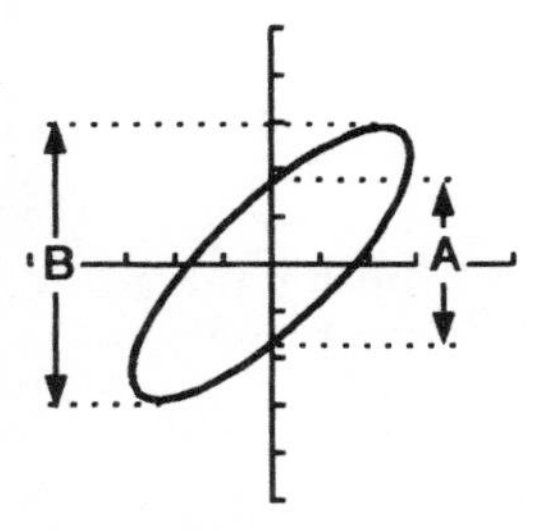

Fig 4.22 Measuring phase from a Lissajous figure.

Phase angle = Arcsin (A/B)

Thus for a value of A/B = 0.5 the angle would be 30°.

One point which needs to be taken into account when measuring the phase difference between two signals is that the two Y amplifiers may themselves introduce a phase difference in the displayed signal. This can be checked by applying the same signal to both inputs, in which case the display should produce a straight line Lissajous figure. This can become important if the test is being carried out at high frequencies.

Modulation test

There are two techniques for measuring the level of modulation of a radio transmitter by using an oscilloscope. Both assume that the Y amplifier has sufficient bandwidth to handle the radio frequency signal produced by the transmitter.

In the first method, the output signal from the transmitter is coupled to the oscilloscope and the level of amplitude modulation can be examined directly. Note that for this measurement the transmitter should be connected to a dummy load and the modulation should be an audio frequency sine wave.

The oscilloscope timebase is set to run at a speed which will display one or two cycles of the audio modulating frequency. The result will be similar to that shown in Figure 4.23.

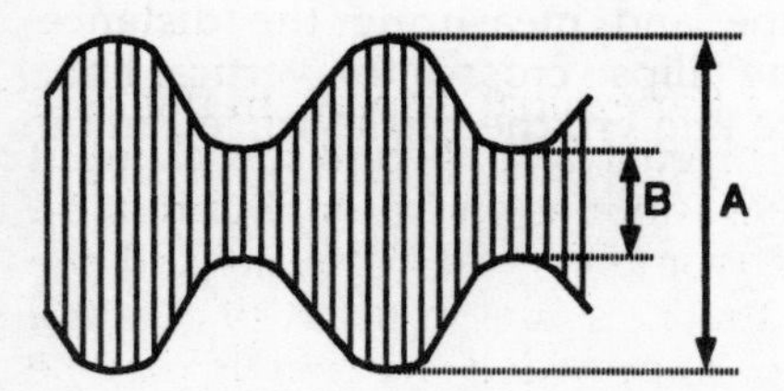

Fig 4.23 Modulation depth using waveform display.

The peak to peak height of the signal A and the height of the unmodulated band B are measured. The modulation percentage is then given by

$$\% \text{ mod} = \frac{A - B}{A + B} \times 100$$

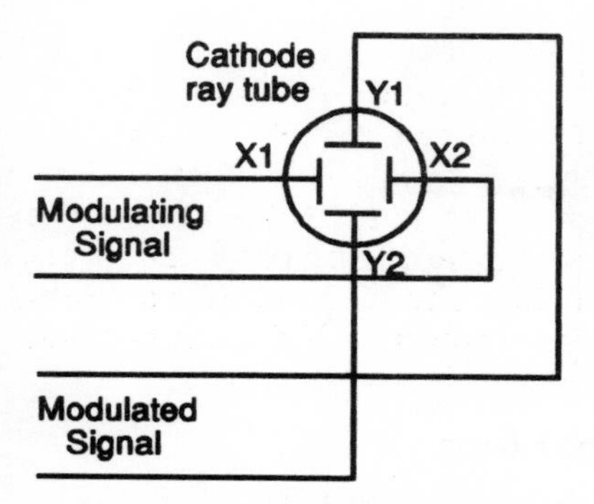

Fig 4.24 Connections for trapezium display.

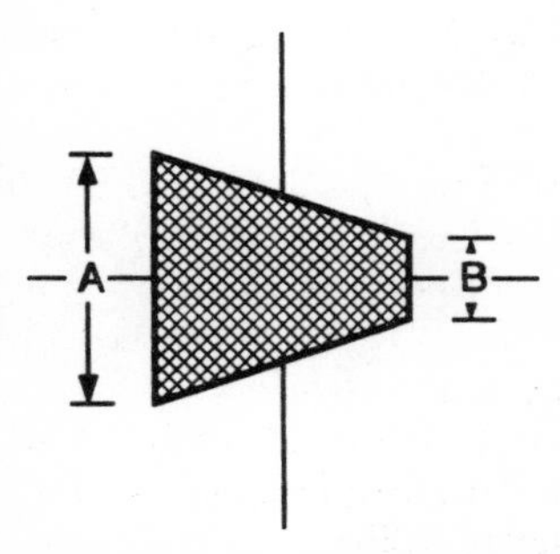

Fig 4.25 Modulation depth using trapezium display.

The alternative method is to use a trapezoidal display. In this case the modulated transmitter output signal is applied to the Y1 input and the modulating signal to the X input as shown in Figure 4.24. In a dual trace oscilloscope this would be the Y2 input and the oscilloscope is switched to pass the Y2 signal to the X plates. If a direct input to the X and Y plates is provided the signals may be applied directly to the plates to avoid any distortion produced by the amplifier circuits.

The display produced is a trapezoidal figure similar to that shown in Figure 4.25. In this case, measurements should be taken of the height of the left side (A) and the height of the right side (B). The modulation percentage is given by

$$\% \text{ modulation} = \frac{A - B}{A + B} \times 100$$

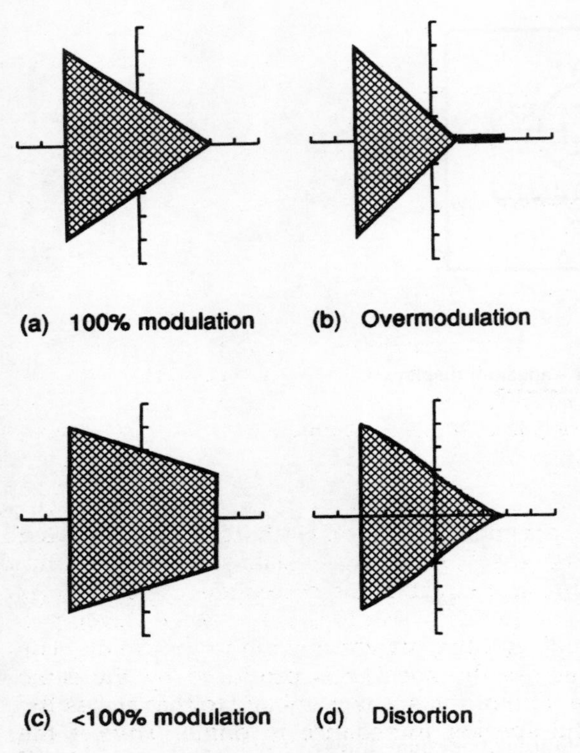

Fig 4.26 Typical trapezium displays.

An unmodulated carrier will produce a vertical line and overmodulation is indicated by a horizontal line at the right end of the trapezoid. Distortion is indicated by curvature of the top and bottom sides of the trapezoid. These effects are shown in Figure 4.26.

Measuring impedances

An oscilloscope can be used to measure the impedance of a circuit by measuring the voltage signal across the circuit and the current signal in a series resistor as shown in Figure 4.27.

In this arrangement a loudspeaker coil impedance is to be measured. The resistor in series might conveniently have a value

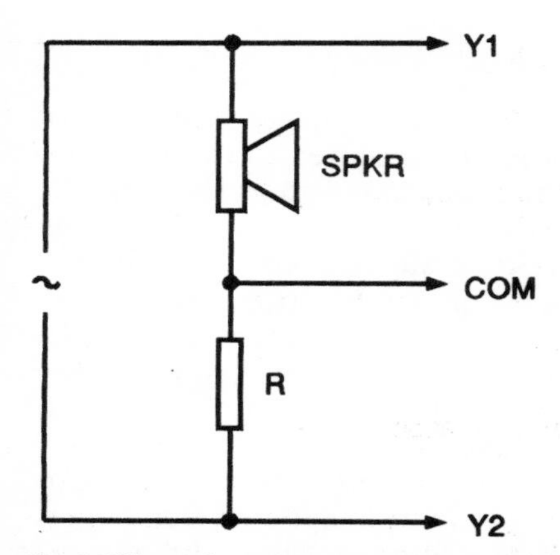

Fig 4.27 Impedance measurement using an oscilloscope.

of 1Ω. Note that the common line of the oscilloscope is connected to the junction of the speaker and the resistor. To energise the circuit a sine wave from a suitable audio frequency generator is applied across the speaker and resistor. The voltage developed across the resistor indicates the current flowing in the circuit. The voltage developed across the speaker is produced by the same series current so the ratio of the speaker voltage to that across the resistor will give the speaker impedance in ohms. Thus if the resistor voltage is 0.5V peak-peak and that across the speaker is 4V peak-peak then the speaker impedance is 8Ω. If a resistor of other than 1Ω is used then the ratio of the two voltages should be multiplied by the resistor value to obtain the speaker impedance.

Normally speaker impedance would be checked at a frequency of 1000Hz. The speaker impedance will vary with frequency and further checks could be carried out at a number of frequencies through the audio range to see how the impedance varies with frequency.

Storage oscilloscopes

The standard oscilloscope is good at displaying repetitive type signals such as sine waves or video signals but is not so good for the examination of transient phenomena or other irregular signals. This type of signal can however be examined by using a storage type oscilloscope. In this type of instrument, a single scan of the timebase is executed and the signal that occurs during that scan is stored and displayed on the screen for a long period or possibly indefinitely so that it can be examined at leisure.

One early technique for examining transient signals involved the use of a long persistence tube for the display. In the normal cathode ray tube used for a standard oscilloscope the image on the screen fades away about 20ms or so after it has been scanned by the electron beam. This can be seen if the timebase is set to scan at perhaps ten sweeps per second when it will be seen that the early part of the trace has already faded by the time the scan reaches the right hand side of the screen.

By using a different screen phosphor material the rate at which the image fades can be reduced so that an image may remain on the screen for perhaps a few seconds. Long persistence tubes were originally developed for use in radar displays and usually had dual phosphor. One phosphor on the screen produces a blue image with a short persistence whilst the second phosphor produces a yellow image which fades out slowly.

Although the use of a long persistence phosphor does allow transients to be examined to some extent, a more ideal scheme is for the image to be retained more or less indefinitely whilst it is examined. This can be achieved by using a special type of tube known as a mesh type storage tube.

Storage tubes

Most of the older models of storage type oscilloscope use a special type of cathode ray tube known as a storage tube. In the basic storage tube a fine mesh screen called the target is mounted a short distance behind the phosphor coating at the screen end of the tube. The mesh itself consists of a fine mesh with 100 to 200 wires per centimetre and is coated, on the side away from the screen, with a layer of insulating material. A short distance away from the insulating layer of the target is another mesh screen called the collector.

The target mesh is held at a potential close to ground and the collector mesh is usually set at a potential of about +100V and its function is to collect secondary electrons emitted from the insulating layer of the target mesh. Between the phosphor coating and the faceplate of the tube is a transparent conducting layer which is held at a potential of some 6kV and this serves to accelerate electrons passing through the target mesh so that they will produce a luminous spot when they hit the screen. The main electron gun of the tube has its cathode set at a voltage of around −1500V and its anode and deflector plates at around ground potential.

The scanning system makes a single horizontal sweep during which the writing of the display trace is carried out. As the beam sweeps over the insulating material of the target it causes secondary electrons to be emitted from the insulating layer at the point where the beam strikes the target mesh. The result is that a series of points on the inner surface of the target insulating layer become positively charged relative to the mesh. Some of the electrons in the beam actually pass through the holes in the target mesh and are accelerated towards the phosphor screen to produce a displayed trace in the usual way.

A second electron gun, called the flood gun, is also used to illuminate the screen through the charged mesh. This gun produces an unfocused beam of electrons travelling at low velocity and during the scan period it has the effect of placing a negative charge on those parts of the target that were not struck by the main beam.

After the writing scan, the negative charge on the mesh blocks the flood beam in those areas where there was no trace during the initial scan but allows the beam through in those places where the writing trace occurred. The flood beam electrons passing through the target mesh are accelerated to the phosphor screen and reinforce the original trace image. The trace therefore remains on the screen for a period of time until the charge on the target mesh insulation layer is lost. Most storage oscilloscopes pulse the beam from the flood gun and this produces much longer storage periods of up to an hour or so. Other improvements in storage tube design have permitted variable persistence displays and improved contrast in the stored image.

Before writing a new trace on the screen the stored trace should be erased. This is achieved by raising the potential of the storage mesh to about +5V for a period of a second. The surface of the insulator attracts electrons from the flood beam and this discharges the stored pattern so that when the erasure pulse is removed the mesh returns to 0V and the surface of the insulating layer becomes negative and this effectively repels the beam from the flood gun thus removing the image from the screen. A new pattern of charge may now be written to the storage mesh by carrying out another writing scan with the main electron gun.

Digital storage

One problem with analogue storage oscilloscopes is that they are very expensive because of the special tube. An alternative technique for providing the facilities of a storage oscilloscope is to use a solid state digital memory system to store the trace information and a conventional oscilloscope tube to produce the display. Virtually all modern storage type oscilloscopes use the digital technique.

The basic arrangement for a simple digital storage oscilloscope is shown in Figure 4.28.

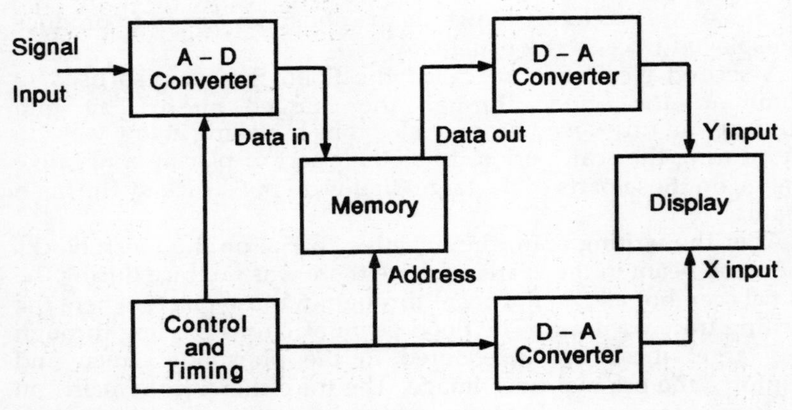

Fig 4.28 Block diagram of digital storage oscilloscope.

The incoming signal is amplified to a suitable level and is then sampled at regular intervals. Each sample level is held temporarily by a sample and hold circuit and the output is applied to an analogue to digital converter (ADC) circuit which produces a binary logic signal corresponding to the amplitude of the input signal at the sampling time. The binary signal is then written into a digital memory circuit where it is stored until required.

In a basic instrument 1024 samples of the input signal might be taken over the period corresponding to one horizontal sweep across the screen. The 1024 digital numbers produced by the ADC would then be written into the digital memory. Each sample value is usually converted into an 8 bit binary number which allows amplitude values from 0 to 255. Thus the memory unit required would have a capacity of 1024 bytes where 1 byte is an 8 bit wide data word. This could be provided in a single integrated circuit memory chip.

Once the data has been written into the memory the circuit can be switched to its reading mode. The data is read out from successive memory locations and passed through a digital to analogue converter to produce an analogue output signal in the form of a series of small steps.

If a conventional X timebase scan is used then the rate at which samples are read from memory has to be adjusted so that the 1024 samples will be read in the time it takes to make one scan across the screen. To provide a stable display the scan must be repeated at least 16 times a second. A synchronisation pulse for the X scan could be derived from the memory address circuits each time a new scan of the memory was started.

A more convenient technique for generating the X sweep signal for the oscilloscope is to make use of the memory address signal. The memory address is a digital number from 0 to 1023 produced by a counter circuit which is incremented at a constant rate. When the address counter reaches 1023 the next clock pulse will reset the counter to 0 to start a new scan of the memory. The address data is fed to a digital-analogue converter and this will produce a sawtooth output signal since the address increases linearly with time. The sawtooth produced by the address signal is then used to drive the X amplifier to provide the horizontal scan and this is automatically synchronised with the data applied to the Y amplifier.

The actual oscilloscope display could be a standard oscillosope and the digital memory plus its control system could be built as an external add on unit. Alternatively the memory system could be built into the oscilloscope to provide storage facilities.

One major advantage of digital storage is that the waveform, once captured in memory, can be displayed on the screen for any desired period of time.

With a resolution of 1024 samples across the screen and 256 levels in the vertical direction, the resultant trace is virtually identical to that produced by a normal analogue signal. The screen space for a single sample would be about 0.3mm vertical by 0.1mm horizontal which is probably smaller than the size of the focused dot on the screen.

To emulate the X expansion facility of a standard analogue oscilloscope, the digital memory could be increased to 8k bytes to allow 8096 samples to be taken for a single horizontal sweep. In the normal display mode all 8096 samples could be displayed to give a very detailed picture. For the expanded mode a block of 1024 successive samples from memory are used to produce the display and the result is that a segment of the full trace is now expanded to fill the screen. By moving the point in memory at which the block of displayed samples starts the displayed segment can be moved through the complete scan. This produces a similar effect to using the X shift control on an analogue oscilloscope when it is in its expanded mode.

Multiple traces can readily be achieved using a digital storage scheme by using a separate bank of memory for each trace and then reading both memories in sequence. A dc offset can be added to the analogue output from each memory unit to separate the traces on the display screen.

Cursor measurements

On most digital storage oscilloscopes an alphanumeric readout showing the current X and Y sensitivity settings is usually displayed on the screen. The oscilloscope may also provide measuring cursors and a readout of the voltage or time settings for the current cursor position.

Some instruments provide two cursors for the Y direction and two for the X direction. One cursor of each pair is a reference cursor which indicates a reference level from which measurements are made whilst the other is a measurement cursor.

For simple level measurements, the two reference cursors might be set at zero reference levels corresponding to perhaps the left of the X axis and the mid-screen point on the Y axis. The X and Y digital level readouts will now indicate the voltage or time levels of the two measurement cursors. As the cursors are moved over the screen by using cursor shift controls the numeric displays change to show the current scale values for each cursor. The cursors can now be moved to various points on the waveform to read out the time or voltage level at that point.

To measure a peak to peak amplitude of a wave the reference cursor could be set to the negative peak and the measuring cursor to the positive peak when the Y level readout will show the peak to peak value in volts.

5 Signal sources

For many of the tests and measurements carried out in radio and electronics work there is a need for some form of signal source. Most applications require the use of a sine wave signal but sometimes other waveforms such as triangular or square waves are more useful. For work involving measurements, or accurate setting up of equipment under test, it should be possible to set the frequency and amplitude of the signal with some degree of accuracy. Work on television or video systems may involve the use of more complex signals which will produce various test patterns on the television screen. In this chapter we shall examine some of the types of instrument which are used to generate signals.

Signal tracer

Perhaps the simplest form of signal source for audio frequency applications is the signal tracer. This is a simple oscillator mounted in a handheld unit with a probe on one end through which the signal may be injected into the circuit under test. A second lead from the signal tracer is clipped to the chassis or ground connection of the equipment under test.

The oscillator in the signal tracer probe is usually an astable multivibrator using a circuit such as that shown in Figure 5.1. This circuit produces a square wave output and a typical frequency for audio applications would be 1000 Hz. Power for the oscillator is usually supplied by one or two dry cell batteries mounted inside the instrument.

Transistors Q1 and Q2 are cross coupled so that, as Q1 turns on, the drop in its collector voltage is transferred to the base of Q2 and this turns off transistor Q2. As Q2 starts to turn off, the rise in its

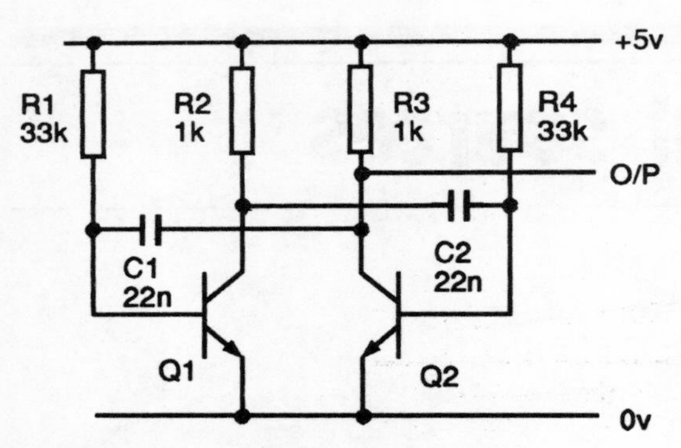

Fig 5.1 Astable multivibrator circuit.

collector voltage is coupled back to the base of Q1 and turns Q1 on. The voltage at the collector of Q1 falls rapidly to almost zero but capacitor C2 is unable to discharge so the base voltage on Q2 goes negative by about 5V.

The timing is governed by the time constant (RC) of the two coupling networks. When transistor Q1 is off its collector voltage is roughly at the +5V supply level. The base voltage of Q2 will be at about 0V and the capacitor C2 will be charged to about 5V. When the circuit switches state the collector voltage of Q1 falls to around 0V but the capacitor is unable to discharge rapidly so the base end goes to a voltage of about −5V which cuts off transistor Q2. At this point capacitor C2 discharges through R4 and the voltage at the base of Q2 rises until Q2 eventually starts to turn on again when the circuit again switches its state.

The time taken for the capacitor C2 to discharge is governed by the values of R4 and C2. The voltage across the capacitor falls exponentially as shown in Figure 5.2 and is given by the equation

$$V_c = V.e^{-t/R.C}$$

where V is the voltage to which the capacitor was originally charged, t is time elapsed. The product of resistance and capacitance R.C is usually referred to as the time constant of the circuit. When elapsed time t is equal to R.C seconds the voltage across the capacitor will have fallen to 0.37 of its original value.

Because the resistor is tied to the +5V line the capacitor can be considered to be initially charged to −10V relative to the +5V line. When the capacitor has discharged by 5V the base of Q2 will be at

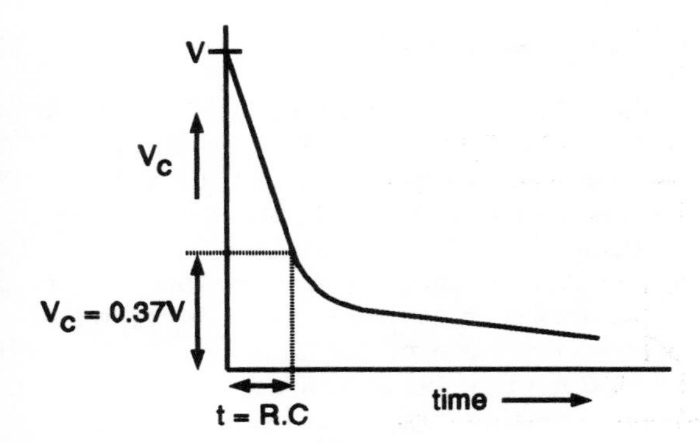

Fig 5.2 Capacitor discharge curve.

0V and the transistor will start to turn on. The time taken for the capacitor to discharge to half its original charge voltage is given approximately by

$$t = 0.7RC$$

and this will be the period for one half cycle of the output waveform. The total period of the square wave generated by this type of oscillator is therefore approximately

$$T = 1.4RC$$

where R and C are the values of the base resistors and cross coupling capacitors. It is assumed here that both coupling circuits have the same values and the square wave produced has an equal duration of half cycles. The actual frequency is modified slightly because the collector voltages of the two transistors do not fall completely to zero volts when the transistor is turned on and the base voltage needs to rise to about +0.3V in order to turn on a transistor.

The oscillator shown generates a square wave of about 1000 Hz which can be used as a simple signal source for checking out an audio amplifier circuit. If the audio amplifier is not working, the signal from this oscillator can be injected into the amplifier at various points starting at the output end of the amplifier and working back to the input. If the injected signal produces an output from the speaker then the part of the amplifier between the probe and the output is working. When no output is obtained from the circuit, the fault is located in that part of the circuit between the point where the probe signal is being injected and the previous test point which gave an output signal.

Although the circuit of this generator is simple the output it produces is a square wave which contains a large harmonic content. Whilst this is useful for checking that an amplifier is working it is not particularly useful for carrying out serious measurement on the performance of the amplifier.

The simple multivibrator type probe can also be useful for checking radio frequency circuits such as those in a radio receiver. The procedure is again similar to that used for an audio amplifier with the probe signal injection point being moved step by step through the receiver circuit starting from the output end and working back towards the antenna input. Because of its high harmonic output there should be a useful level of signal extending well into the radio frequency bands. When fed into the antenna input of a receiver the multivibrator should produce signals at

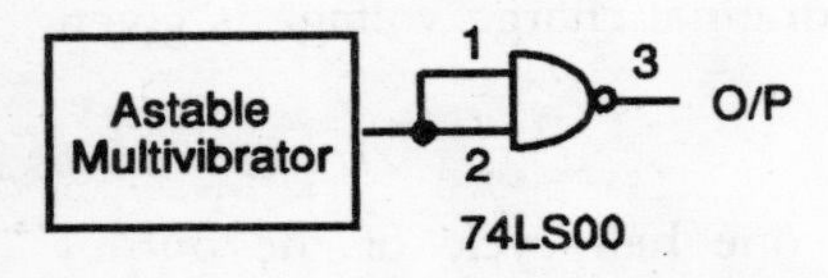

Fig 5.3 Use of Schottky logic buffer output.

1kHz intervals throughout the receiver frequency range. To provide better output at the higher radio frequencies a Schottky TTL gate can be added as an output driver stage as shown in Figure 5.3. This type of gate should be able to produce harmonics up to frequencies of 50MHz or higher.

Audio oscillators

So far we have been looking at simple single frequency oscillators with square wave output. Whilst these are useful for signal tracing and for frequency calibration they are not so useful for measurements of the performance of an audio amplifier or a radio receiver.

The most convenient signal for carrying out measurements in an audio circuit is a pure sine wave at a known frequency. This type of signal can be generated by using an amplifier with a frequency selective resistor-capacitor (RC) network connected so that it feeds part of the output signal back to the input. If we now arrange that at one particular frequency, determined by the RC network, the signal fed back produces an in-phase signal at the output of the amplifier, the circuit will oscillate at that frequency.

A typical audio frequency signal generator should be able to produce frequencies from about 20Hz up to perhaps 20 kHz with signal levels up to one or two volts RMS.

Phase shift oscillator

One simple type of circuit for producing audio frequency signals is the phase shift oscillator shown in Figure 5.4. In this circuit the

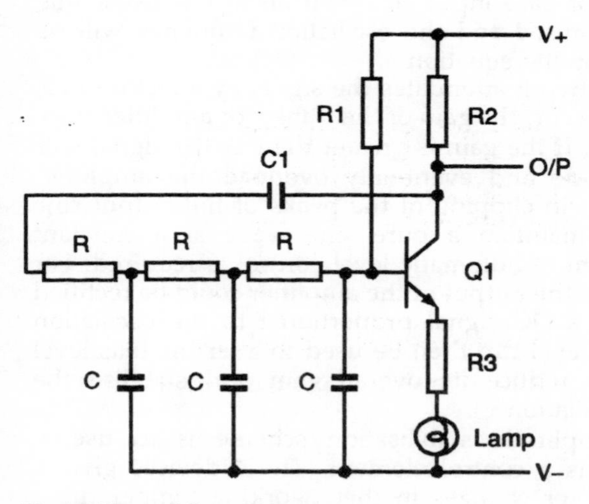

Fig 5.4 Phase shift oscillator circuit.

output of the amplifier is fed to a three stage RC network in which each section has the same values for R and C. If we consider one RC section and feed an AC signal to it, the output signal across the capacitor will lag in phase behind the input, and the amount of phase shift will increase with frequency. At some frequency, the phase shift produced by each section of the network becomes 60 degrees so that the total phase shift through the three sections of the network is 180 degrees. If we now arrange that the amplifier also produces a 180 degree phase shift then the signal fed back to the input will reinforce any signal that is already being input and, provided there is sufficient gain in the amplifier, the circuit will oscillate. To achieve 180 degrees phase shift in the amplifier we need a simple inversion of the signal which can be obtained by using a single stage amplifier.

The frequency at which the circuit will oscillate is approximately

$$f = \frac{10^6}{15.4\ RC}\ \text{Hz}$$

where R and C are the values of the resistor and capacitor in each section of the phase shift network. In this equation R is in ohms and C is in microfarads. The actual frequency of oscillation will be slightly different from this theoretical value due to the loading effect of the transistor base input circuit. If an FET is used, this loading will be minimised and the oscillation frequency will be closer to that given by the equation.

The phase shift network attenuates the signal by a factor of 29, so, for oscillation to occur, the gain of the transistor amplifier stage needs to be set at 29. If the gain is greater than 29 the signal will build up in amplitude and eventually overload the amplifier, which usually results in clipping of the peaks of the output sine wave. In order to maintain a pure sine wave at a constant amplitude, some form of automatic level control is required. For this the signal level at the output of the amplifier could be rectified and filtered to give a DC signal proportional to the oscillation amplitude. This DC level can then be used to alter the bias level on the amplifier to reduce its overall gain and stabilise the amplitude of the oscillation.

An alternative amplitude stabilisation scheme is to use a thermistor or lamp as a control element. These devices give a change in resistance for changes in their working temperature. Since the working temperature is governed by the current flowing through the lamp or thermistor the resistance will be controlled by the voltage applied across the device. For a lamp, the resistance rises with temperature so the lamp could be inserted in series with the emitter of the transistor to provide negative feedback. As the oscillation amplitude rises the lamp takes more current and increases its resistance to reduce the overall gain of the amplifier. The characteristics of the lamp or thermistor used will need to be matched to the gain characteristics of the amplifier circuit for proper operation.

For a single frequency generator the phase shift circuit is quite effective. If we want to vary the output frequency however, a problem arises. To change the frequency of operation, the value of R or C in each section of the network must be changed. The simplest method of frequency control is to change the value of the three resistors in the feedback network. To maintain proper operation, all three resistors need to be changed simultaneously and this means using a three gang potentiometer as the frequency control.

Wien bridge oscillator

The need for a three gang potentiometer to provide a frequency control in the phase shift oscillator makes this circuit unattractive for use as a general purpose audio signal generator. An alternative oscillator which provides a more practical solution is the Wien bridge type which has the basic circuit arrangement shown in Figure 5.5.

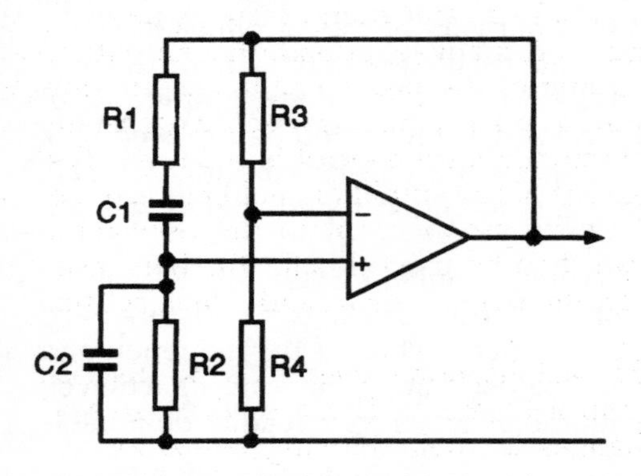

Fig 5.5 Wien bridge oscillator circuit.

In this circuit the output of the amplifier is applied across the input terminals of a Wien bridge circuit. When the bridge is balanced, the phase shift produced by R1 C1 is exactly balanced by an equal but opposite phase shift produced by R2 C2, and the signal at the junction of R1 and R2 is in phase with the signal input across the bridge. If R1 = R2 and C1 = C2 then the impedance of R2 and C2 in parallel is equal to half the impedance of R1 and C1 in series so the output amplitude produced across R2 C2 is one third of the amplitude of the signal applied to the bridge. If R3 is made equal to 2R4 then the signal at the junction of these resistors is exactly the same as that at the junction of R1 and R2 and the bridge becomes completely balanced. With a fixed set of values for R1, R2, C1 and C2 there is only one frequency at which bridge balance occurs. If R1 and R2 are both equal to a value R with C1 and C2 having a capacitance C then the frequency for balance is

$$f = \frac{10^6}{6.28\ RC}\ \text{Hz}$$

where R is in ohms and C is in microfarads.

The output signal across R2 is fed to the non-inverting input of an operational amplifier and the output of the amplifier is used to drive the bridge. This provides positive feedback from the amplifier output via the bridge to the amplifier input. The gain of the amplifier is controlled by the potential divider R3, R4 which forms the other arm of the bridge. The signal across R4 is fed to the inverting input of the amplifier to provide negative feedback, and if R3 = 2 x R4 then the overall gain of the amplifier becomes 3 which is just enough to maintain oscillation in the circuit. A small adjustment of the ratio of R4 to R3 allows the amplitude of oscillation to be controlled. The resistor R3 could be replaced by a thermistor and if the characteristics are chosen carefully a stable amplitude of oscillation can be maintained together with a good sine wave output.

In the Wien bridge circuit there are only two resistors R1 and R2 which affect the frequency of oscillation. These two resistors need to be changed in unison to produce a change in frequency and maintain proper operation of the oscillator. This can be achieved by using a twin gang potentiometer which is a readily obtainable component.

In an audio frequency signal generator, the output from the Wien bridge oscillator would be fed to a buffer amplifier which provides the drive for the output terminals of the instrument. The buffer amplifier isolates the oscillator circuit from any loading effects caused by the circuit under test so that the output level and waveform remain constant. The output from the buffer amplifier is then passed through an attenuator circuit to the output terminals of the instrument. In a low price generator, the attenuator is simply a potentiometer which permits the output signal level to be adjusted. More advanced instruments are usually fitted with digital step attenuators to permit the setting of known output levels.

Waveform generators

Most modern audio signal sources provide not only a sine wave output but also square and triangular wave signals as well. These instruments are generally referred to as waveform generators to distinguish them from the ordinary signal generators which produce only a sine wave output.

In these instruments, the basic triangular waveform is generated first by using a capacitor charged and discharged at constant current as the timing device. The basic block diagram for such an instrument is shown in Figure 5.6. The triangular signal is generated by using the voltage produced across a capacitor which is charged and discharged alternately by being switched to two constant current generators I1 and I2. The voltage across the capacitor is fed to a pair of level comparators which detect when

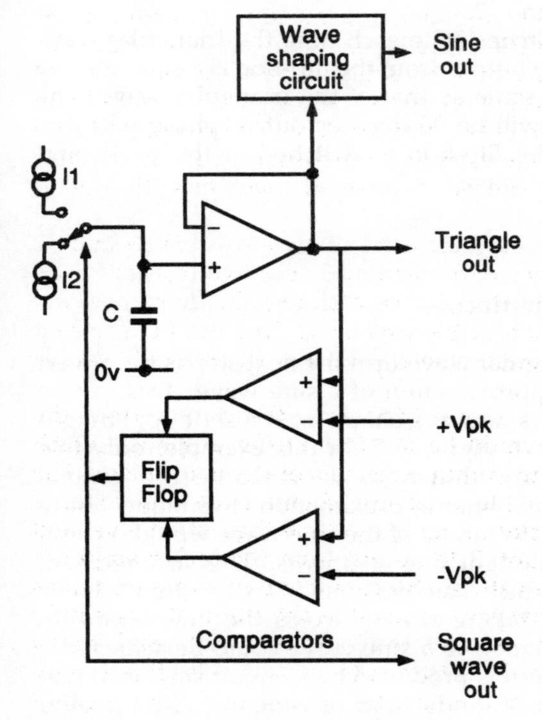

Fig 5.6 Waveform generator block diagram.

the capacitor voltage reaches two preset voltage levels. The output of the comparators drive a flip-flop which in turn switches the constant current sources I1 and I2 via switch S1.

For the rising slope of the triangular wave the capacitor is switched so that it charges linearly with time from current source I1. When the capacitor voltage reaches the reference level of comparator A1 the output of A1 triggers the flip-flop circuit and this in turn operates switch S1. The capacitor is now discharged by the current source I2 and falls linearly with time until it reaches the reference level of comparator A2. The output of A2 is used to reset the flip-flop FF and this operates switch S1 so that the capacitor again charges from I1 to start a new oscillation cycle. The result is that the voltage across the capacitor rises and falls linearly between the two reference levels to produce a triangular output waveform. The amplitude of the waveform is determined by the voltage reference levels applied to the two comparators and the frequency by the value of the capacitor and the levels of current from the generators I1 and I2.

Since the flip-flop switches state each time the triangular wave reverses its direction, the output from the flip-flop is a square wave whose frequency is the same as that of the triangular wave. The square wave produced will be 90 degrees out of phase with the triangular wave since the flip-flop is switched at the peaks and troughs of the triangular wave.

Sine wave shaping circuits

Having produced a triangular waveform the next step is to convert this into a reasonable approximation of a sine wave. This can be done by building up the sine wave shape from a series of straight line segments as shown in Figure 5.7. In this example, only five straight line segments are shown to demonstrate the general principle of the straight line approximation technique. In a practical circuit, each quarter cycle of the sine wave would be built up from a set of five straight line segments. This arrangement produces a sine wave which has less than 1% distortion relative to a pure sine wave. By taking one or two further line segments an even better approximation to a sine curve could be achieved.

The wave shaping action is produced by a circuit similar to that of Figure 5.8. Here a set of diodes and resistors are used to alter the shape of the positive half cycle of a triangular wave to produce an approximation to the positive half cycle of a sine wave.

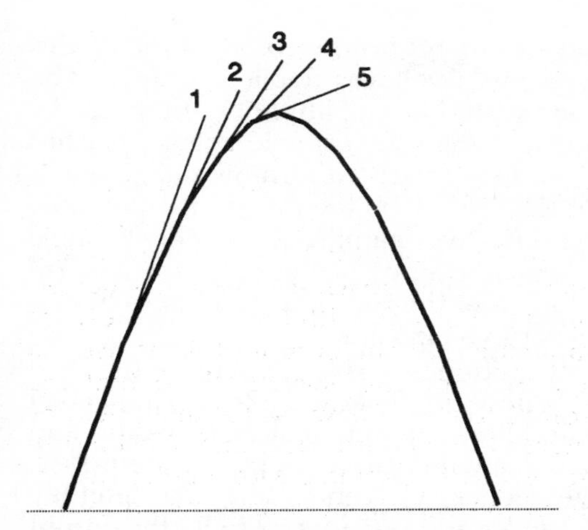

Fig 5.7 Sine wave built from straight line segments.

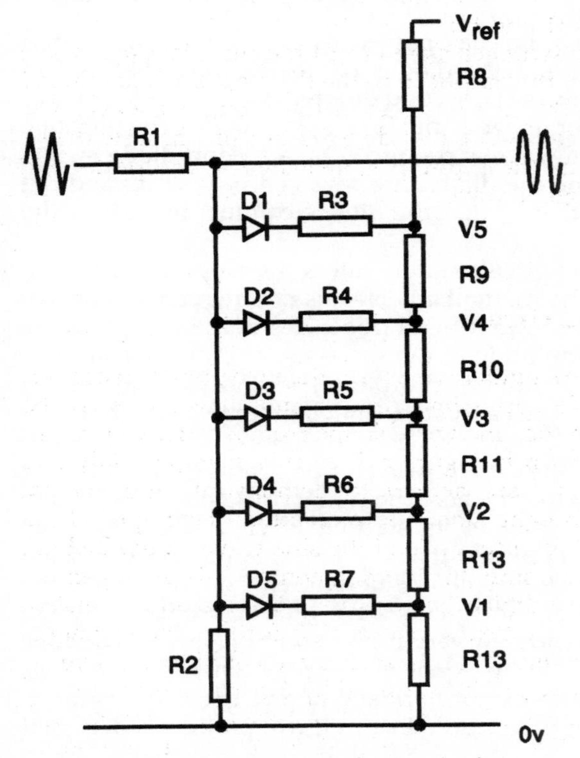

Fig 5.8 Triangle to sine waveshaping circuit.

When the triangular wave starts off from zero, all of the diodes are turned off by the reverse bias voltages applied to them. The triangular wave input is attenuated in amplitude by the potential divider formed by R1 and R2. When the triangle voltage reaches the first bias level V1 the diode D5 starts to turn on and connects R7 in parallel with resistor R2. This alters the ratio of the potential divider and reduces the effective amplitude of the triangle waveform and from this point the slope of the output waveform is reduced. It is assumed here that the bias voltage sources have a very low resistance to the zero voltage line.

As the level of the output continues to rise it will reach the level of bias voltage V2. Diode D4 now conducts and adds R6 in parallel with the output. This again alters the potential divider ratio and the output signal takes up another new slope. This process continues until all of the diodes are conducting. By carefully adjusting the bias levels V1 to V5 and resistors R3 to R7 the output voltage can be made to follow the curve for a quarter cycle of a sine wave.

After the triangle wave has passed its peak level, the output voltage starts to fall and the diodes D1 to D5 will switch off in reverse order to produce the second quarter cycle of the sine wave which will be a mirror image of the first quarter cycle. The negative half cycle of the sine wave is produced by a second diode switch network but this time the diodes are reversed in polarity and the bias levels are negative so that the diodes conduct in turn as the signal goes negative.

Because the shaper circuit simply alters the slope according to the voltage level of the triangular wave it is not frequency sensitive and will produce a sine wave at all freqencies of the triangle wave provided that the amplitude of the drive signal remains constant.

Some waveform generators allow the rise and fall rates of the triangular wave to be adjusted independently so that the waveshape can be altered from an equal slope triangular wave through various intermediate waveforms until eventually the signal becomes a sawtooth where the signal rises linearly with time and then is rapidly reset to its starting level. If the amplitude remains the same this means that the time periods for the rising slope and the falling slope will be different which in turn alters the mark to space ratio of the square wave output produced by the flip-flop. This signal can now vary from a short positive or negative going pulse through to an equal mark-space square wave. This mark to space variation is not normally of any great use as far as the sine wave output is concerned since it will produce a distorted sine wave output where the positive and negative half cycles are not symmetrical.

The ICL8038 waveform generator IC

A very useful and versatile audio frequency waveform generator can be built using the GE/Intersil ICL8038 integrated circuit which is capable of generating sine, triangle and square wave outputs simultaneously. This chip contains the constant current generators, comparators and flip-flop for generating the triangular wave as well as the diode shaping circuits for converting the triangle into a sine wave. By adding a few external resistors and capacitors and a buffer amplifier a versatile waveform generator can be constructed.

Figure 5.9 shows the pin connections of the 8038 integrated which comes in a 14 pin DIL package. The 8038 needs a split

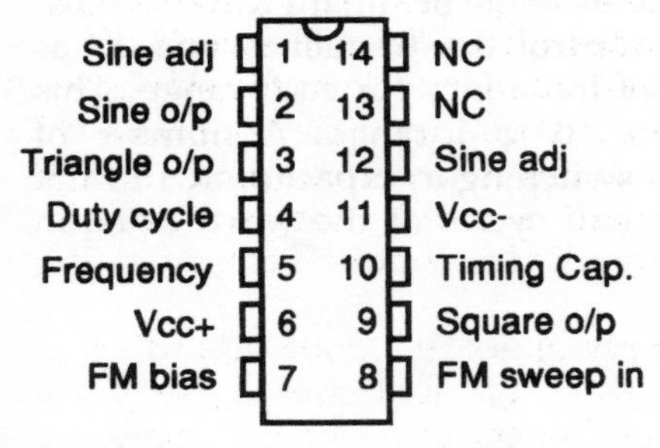

Fig 5.9 Pin connections of ICL8038 integrated circuit.

supply which might typically be +10V and −10V but the device will run on supplies down to +5V and −5V. When operated from 10V supplies, the 8038 will produce about 4V peak to peak from its sine wave output and about 6.5V peak to peak from the triangular wave output. The square wave output is a much larger signal with a typical value of 16V peak to peak.

Pins 4 and 5 provide control of the charge and discharge current levels and allow the duty cycle of the basic triangle waveform to be adjusted. Pins 1 and 12 control the bias levels in the two shaper circuits that generate the sine wave output. By adjusting the voltage levels to these pins, the shape of the sine wave can be adjusted for minimum distortion. The voltage applied to pin 8 controls the frequency in conjunction with the capacitor connected between pin 10 and the 0V line. By applying a varying voltage to pin 8, frequency modulation can be achieved. Thus by applying a sawtooth signal to pin 8 the 8038 can be used as a sweep generator.

The output of the wave shaping circuit is not really suitable for use as a direct output from the instrument since the waveshape could be affected by loading caused by external circuits. A buffer amplifier is used to provide the actual output signal. This amplifier can be a 741 type provided that the output frequency does not exceed about 10kHz. The 741 has a relatively low slew rate for its output stage and will be unable to follow the high frequency signals which causes reduced amplitude on sine and triangle waves and rounding of the edges on a square wave. By using an LF351 which has a high slew rate output the frequency range can be extended up to about 1MHz.

Figure 5.10 shows the circuit for a multi range oscillator based on the use of an 8038 waveform generator. Here the ranges can be adjusted to provide a ten to one frequency range and the output can be switched to give either sine, triangle or square wave signals.

Potentiometer R2 is used to control the frequency and R1 is preset to give the desired span of frequency for each range. This might typically be set to give a 10 to 1 range. A number of frequency ranges are selected by switching in capacitors C1 to C5. R5 is adjusted to make the two half cycles of the wave equal in

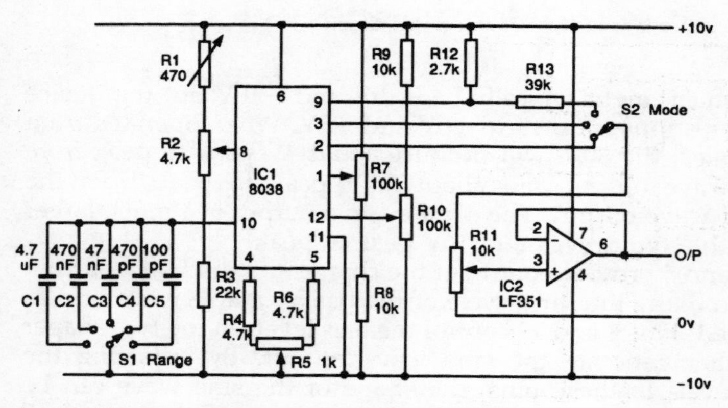

Fig 5.10 Waveform generator circuit using 8038 IC.

duration whilst R7 and R10 are adjusted to alter the bias levels on the diode networks for minimum distortion on the sine wave output. R11 controls the level of the output signal.

Radio frequency generators

Applications such as radio receiver alignment and calibration generally require the use of a radio frequency signal generator where the frequency can be set accurately to any value within the generator's frequency range. The basic arrangement of a typical RF signal generator consists of a simple LC tuned oscillator followed by a buffer amplifier which isolates the oscillator from any loading effects produced by the circuit which is being tested. In addition there will usually be facilities for applying modulation to the signal.

The oscillator circuit can be a simple Hartley type osillator similar to that shown in Figure 5.11. Tuning is carried out by a variable air spaced tuning capacitor, and a number of different inductors may be switched into circuit to give a series of tuning ranges. The main disadvantage of the Hartley type circuit is that in order to change the frequency range a two pole switch is required with one pole selecting the inductor and the second pole selecting the tap point on the new inductor.

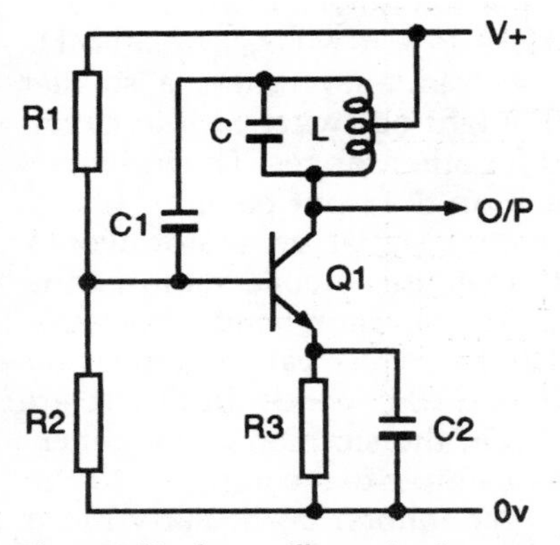

Fig 5.11 Hartley oscillator circuit.

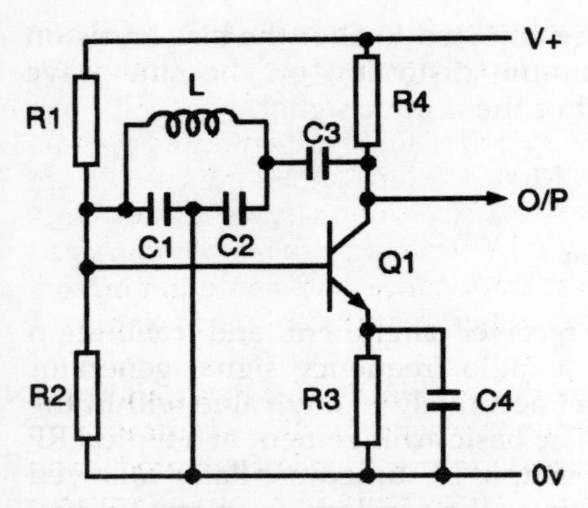

Fig 5.12 Colpitts oscillator circuit.

An alternative circuit which overcomes the need for tapped inductors is the Colpitts type oscillator shown in Figure 5.12. In this oscillator the capacitor circuit is tapped.

Variations of these two basic oscillator circuits are often employed in order to achieve improved frequency stability. The actual oscillator circuit is normally enclosed in a screened box within the instrument and in some cases the internal temperature of the oscillator section may be controlled at a constant level to reduce problems with frequency drift.

A typical generator for use in a radio/TV repair shop might cover the frequency range from 100 kHz up to perhaps 50 MHz. With a fairly standard 500 pF tuning capacitor the ranges might be 100 – 300 kHz, 300 kHz – 1 MHz, 1 – 3 MHz, 3 – 10 MHz, 10 – 30 MHz and 30 – 50 MHz. For the highest frequency range, a smaller tuning capacitor of perhaps 100 pF might be switched into circuit instead of the 500 pF capacitor used for other ranges. The inductors are switched by a band select switch on the front panel.

For VHF and UHF work a separate signal generator would normally be used and this might cover the mobile radio bands between 70 and 200 MHz and the UHF television band. Professional signal generators intended for laboratory use can often provide complete coverage of MF, HF, VHF and UHF bands, but these are much more expensive instruments. On the simpler signal generators the frequency is read off from a simple dial connected to the tuning capacitor. In many modern generators, particularly those intended for laboratory use, a digital readout of frequency is also provided. This is generally obtained from a built in digital frequency counter and gives a more accurate readout than the simple analogue dial.

The output of a signal generator is usually designed to operate into either a 70Ω or 50Ω load. Most modern signal generators are designed for 50Ω operation which is the generally used industrial standard. FM radio and domestic TV receivers are usually matched for operation from a 70Ω input impedance and some generators intended for general radio/TV servicing applications may have a 70Ω output impedance.

On the simpler RF generators the amplitude of the output signal is controlled by a simple potentiometer and no direct indication of the output level is provided apart from a simple calibrated knob on the output level control. Some types include a step attenuator after the potentiometer. In this case it is possible to accurately change the signal level in 10dB or 20dB steps which can be useful for some types of receiver measurements. A generator intended for laboratory use produces a precisely defined output voltage which is fed to the output terminals via a calibrated step attenuator. With these instruments the output voltage can be set to any desired level by adjusting the setting of the attenuator. In some cases, a built in meter provides a direct readout of the generated voltage at the attenuator input.

Amplitude modulation

A broadcast band radio receiver covering long, medium and short waves is normally designed to receive amplitude modulated signals. Applying a constant level signal to such a receiver will generally produce no audio output from the speaker. It is possible to detect a DC level change in the detector circuit of the receiver and this could be monitored using an oscilloscope. To check the overall operation of the receiver, we need a signal which is amplitude modulated by an audio frequency tone. Signal generators usually include facilities for generating an amplitude modulated signal for receiver tests. The audio frequency tone is usually at 1000 Hz. The level of modulation may be fixed at about 80% on simple generators but many instruments allow the modulation depth to be adjusted to any desired level.

Most RF signal generators include a simple 1 kHz audio frequency generator to produce a sine wave signal which is then used to amplitude modulate the RF output. The modulation process causes the amplitude of the RF signal to be varied in sympathy with the audio signal.

The degree of modulation is expressed as a percentage figure. If the modulation has the same amplitude as the RF carrier signal then the instantaneous amplitude of the RF signal will vary from zero up to twice the normal output level and this state is 100% modulation.

The modulator circuit used may be the simple circuit shown in Figure 5.13. Here the supply voltage to the output stage of the RF

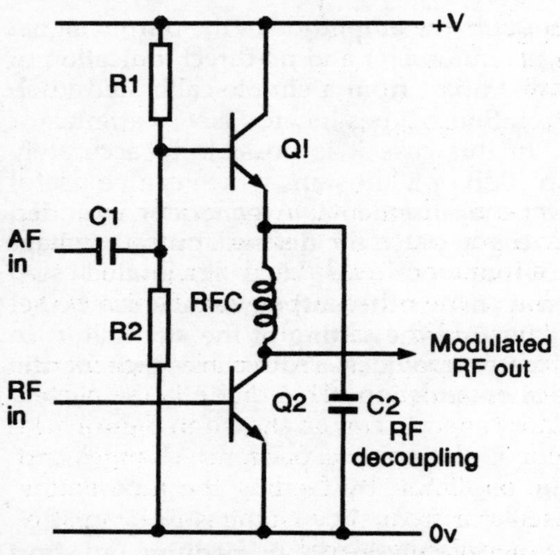

Fig 5.13 Simple amplitude modulator circuit.

generator is varied at audio frequency by the modulator stage Q1. If the modulation is applied before the signal reaches the output amplifier then it is important that the output amplifier operates in a linear mode otherwise distortion of the modulated signal will occur.

In a practical signal generator the modulation level is usually set at around 70%. This provides an adequate audio output signal without causing distortion to the RF signal or the audio component.

Frequency modulation

Radio receivers for VHF broadcast reception are designed to receive frequency modulated signals in which the carrier frequency

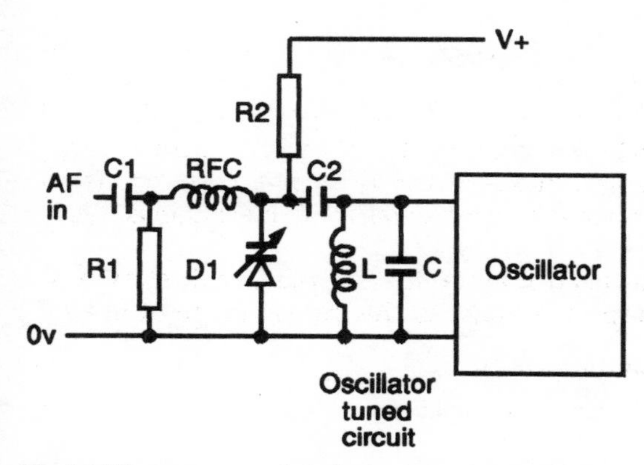

Fig 5.14 Frequency modulation using a varicap diode.

is varied on each side of a centre frequency in sympathy with the modulating signal. A signal generator for testing such a receiver should be able to provide a frequency modulated signal which has similar characteristics to that of the broadcast signal. Typically the modulating signal might be a 1 kHz audio tone.

Modern signal generators normally use a variable capacitance diode to provide frequency modulation. This diode is connected in parallel with the oscillator tuned circuit as shown in Figure 5.14 and as the bias on the diode is altered its capacitance changes and alters the frequency of the oscillator. By feeding the modulating signal to the diode the oscillator frequency changes in sympathy with the waveform of the modulating signal.

Sweep generators

For alignment of receiver IF circuits it is useful to be able to examine the frequency response of the IF amplifier. Normally this frequency response is a bandpass response centred around the nominal value of the intermediate frequency. Thus for an MF receiver the IF is usually 455 kHz and the passband extends perhaps 6 – 10 kHz each side of this frequency. The optimum response of the amplifier will be a curve something like that shown in Figure 5.15.

It is possible to plot the frequency response of the IF amplifier by setting the signal generator to a number of frequencies through the IF response range and measuring the detector output voltage

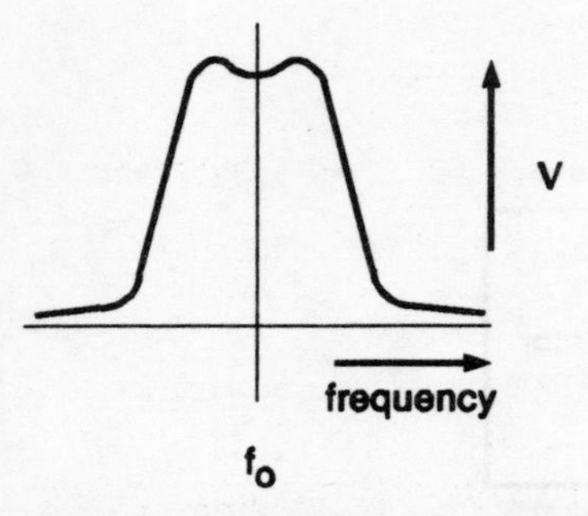

Fig 5.15 Frequency response curve for IF amplifier.

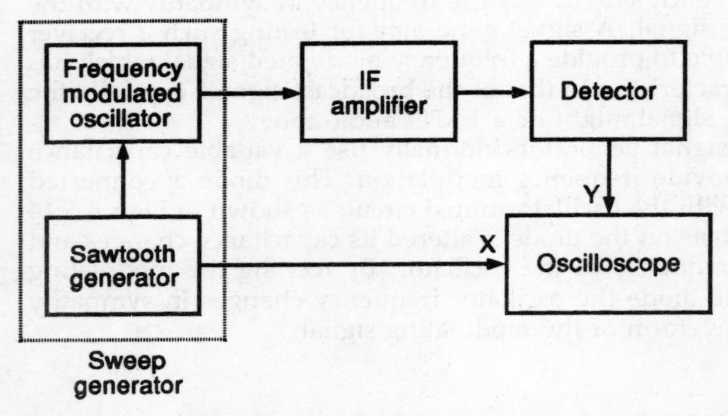

Fig 5.16 Sweep generator set up for IF response check.

at each frequency. This is rather a lengthy and tedious process and a more convenient scheme is to use a sweep frequency generator and an oscilloscope to display the response curve directly.

The basic arrangement for testing the response of an IF amplifier is shown in Figure 5.16. A special signal generator, known as a sweep generator is used to produce the input signal for the IF amplifier. This generator produces a frequency modulated output signal but the modulation is in the form of a sawtooth wave. Thus the frequency is swept linearly through the IF response range and at the end of each sweep, returns immediately to the starting frequency. The sweep rate might be perhaps 50 sweeps per second.

The sawtooth signal from the generator is used to drive the X scan on an oscilloscope and the Y input to the oscilloscope is taken from the receiver detector. As the signal frequency sweeps through the IF response range, the output of the detector will follow the response curve of the circuit under test and this will be displayed on the oscilloscope screen.

The advantage of using a sweep generator and oscilloscope is that as each adjustment is made to the IF tuned circuits the result is displayed immediately on the oscilloscope screen. This same set up can also be used for an FM receiver system to display the response of the FM discriminator circuit. To display the IF response for an FM receiver the oscilloscope might be connected to the AGC output from the IF amplifier, or alternatively a diode detector might be added to allow the IF signal amplitude to be measured. Since most FM receivers include amplitude limiter stages it is important to use a low input signal level so that the limiter stage is not driven hard enough to limit the signal amplitude.

A sweep generator is usually essential for aligning older TV receivers which use tuned circuits to define the IF response. Most modern TV receivers use surface wave filter circuits to produce the correct IF response and these need no adjustment. The only alignment required for such receivers is to set up the tuning of the synchronous detector of the video IF which can be achieved by using a simple signal generator.

A sweep generator operating in the audio frequency range can be used to check the frequency response of an audio amplifier. For this, the set up is similar to that for a receiver. The sweep generator is fed to the audio amplifier input, and the output of the audio amplifier is fed to a detector circuit to produce a peak reading of the audio output voltage. The sweep rate in this case might be reduced to perhaps 10 sweeps per second. In order to produce a wide sweep range the generator in this case will usually produce a radio frequency signal initially at perhaps a few megahertz. The frequency is then swept by perhaps 20kHz which is a small shift when the basic frequency is a few megahertz. The swept signal is then mixed with a fixed frequency equal to the lowest frequency of the swept signal. The output of the mixer will be the original signals together with sum and difference frequencies. The difference frequency in this case will sweep from zero up to 20kHz and by using a low pass filter, this signal can be extracted from the mixer output.

The response of audio frequency filters can also be examined using a sweep generator. This can be useful if the filter is to be fine tuned to give a sharp notch bandstop characteristic or where several stages are being used to produce a flat bandpass filter.

TV pattern generators

For alignment and testing of television receivers, a pattern generator can be a useful instrument. This type of instrument generates a signal which produces one of a number of different types of display on the television screen. The patterns provided range from simple dot or crosshatch patterns, through grey scale and colour bar signals to full scale test patterns similar to those broadcast by TV stations outside programme hours.

Some types of pattern generator are designed to produce one specific type of test signal such as a cross hatch pattern or a colour bar. Most of the modern TV test instruments can produce several types of test pattern. Instruments of this type usually provide both a video signal output and a signal modulated on to a UHF carrier which can be injected into the antenna terminals of the receiver.

Crosshatch generator

For television servicing, a useful instrument is the crosshatch pattern generator. This instrument is designed to generate a video signal which produces a grid of horizontal and vertical lines on the screen of the television set under test. The general idea of this type of screen display is shown in Figure 5.17.

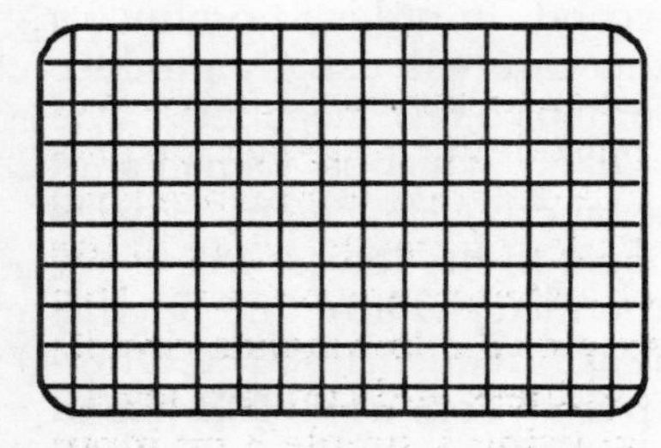

Fig 5.17 Typical crosshatch pattern for TV tests.

The crosshatch pattern is useful when setting up the horizontal and vertical scan circuits of the television receiver. Since the lines on the displayed grid are equally spaced, the pattern can give a direct visual indication of any non-linearity in either horizontal or vertical scan, and adjustments can be made to the scan circuits to produce the best linearity on the display. The pattern is generally chosen so that it produces true squares when the picture aspect ratio is correct and it can be used to set up the appropriate amplitudes for vertical and horizontal scans to give a correctly proportioned picture on the screen.

In a colour receiver, the crosshatch pattern can be used for setting up the convergence circuits which ensure correct alignment of the three electron beams to give proper colour registration in the displayed picture. The main advantage of the pattern over a normal programme picture is that the pattern is static and provides clear horizontal and vertical reference lines. With the older delta gun tubes where the setting up of the convergence is relatively complex, a cross hatch generator is almost essential if correct convergence is to be achieved. Test card patterns are broadcast by many of the TV channels but this is usually early in the morning before programme transmissions commence. With the tendency for stations to change to 24 hour broadcasting the test card broadcasts are becoming very infrequent and therefore of little use to the service engineer.

Alignment of convergence with the more modern in line tubes is a much simpler process but a crosshatch generator can still be useful for checking convergence on this type of tube.

A variation of the crosshatch pattern which is sometimes provided is a pattern of dots. For this type of display a dot is generated at each of the crossing points between the vertical and horizontal lines of the crosshatch pattern.

Grey scale generator

A useful signal for setting up the luminance channel in a receiver is a grey scale pattern. This consists of a series of wide vertical bars across the screen and each bar is set to a different luminance level ranging from black through various shades of grey to white. If the signal is examined it takes the form of a staircase waveform. This signal can be used to check the linearity of the luminance circuits and also used to set up the correct brightness and contrast levels.

The grey scale bars are generated by using a simple 3 bit input digital to analogue converter which produces eight analogue output levels. The D to A converter is driven by a simple three stage binary counter which is clocked by a signal running at about 9 times the line scan frequency. The analogue output signal is then scaled to give a luminance signal of approximately 0.66V peak amplitude. This luminance signal is combined with synchronisation pulses and blanking signals to produce a 1V video output signal.

Colour bar generator

Another useful type of pattern generator for use in television servicing is the colour bar generator. This device generates a display which consists of a series of broad vertical bars on the screen where each bar is displayed in a different colour. The usual sequence of colours is black, blue, red, green, magenta, cyan, yellow and white. This sequence is in fact the set of combinations of the three colours red, green and blue and is built up as follows:

White	red, blue and green
Yellow	red and green
Cyan	blue and green
Magenta	blue and red
Green	green only
Red	red only
Blue	blue only
Black	no colour at all

The cyan colour is a pale blue green colour. If the luminance signal is examined this will be seen to be a staircase with the lowest level at the black end of the bar and the highest at the white end of the bar. The composite video signal has a similar waveform except that there will be bands of colour subcarrier superimposed on the luminance levels for all bars except the black and white ones.

It is possible to generate a simple colour bar signal by using a home computer as the signal source. Assuming that the computer is capable of colour displays, and most home computers are, the main requirement is that the computer should be able to display the eight basic colours of the colour bar pattern. The simpler home computers, such as the BBC Micro, Spectrum, Amstrad CPC and Commodore 64, do in fact use the 8 basic combinations of red, green and blue to provide the set of display colours. More advanced computers such as the Amiga and Atari ST are capable of providing a much wider range of display colours but can readily be programmed to generate the set of colours needed for a colour bar display. In all cases a simple program in BASIC can be used to generate the desired colour bar display, and for most computers there will be a modulated UHF output signal available which can be fed to the antenna input socket of the television receiver under test.

Test card generators

For testing and aligning television and video systems, a test card generator can be a useful instrument. The test card display typically includes the functions of a crosshatch pattern for picture linearity adjustments, a colour bar pattern for checking the alignment of the chrominance circuit, a grey scale pattern for checking luminance circuit adjustments, and resolution patterns for checking the bandwidth of the amplifier and rf circuits. Most test cards also include a circle pattern which is a convenient display for checking the aspect ratio of the displayed picture.

A variety of test card patterns have been used by the broadcast companies during transmission periods outside normal programme hours. In Britain, many stations now operate programming 24 hours a day and others are only off the air for a few hours early in the morning so that test cards are rarely broadcast at times where they would be useful for servicing purposes. For convenience in servicing work on television receivers or video systems, a test card generator can be a useful instrument.

Most modern test card generators use electronic logic systems to generate the required screen display. The basic screen pattern is usually a cross hatch type display. This signal is gated off over various portions of the display area and other signals are inserted into these areas of the picture. If a circle pattern is included, this can usually be added in by simply OR gating it with the other video signals. A colour bar signal is usually included in the display and a luminance grey scale bar can also be included. For bandwidth checking a third horizontal bar with a number of bursts of high frequency square wave signals is included. Typically the frequency bars might be at frequencies of 0.5, 1.0, 2.0, 3.0, 4.0 and 5.0 MHz. These signals will often be derived from the quartz crystal used to generate the colour subcarrier. The sync pulses are also derived from these signals so that the frequency in each of the resolution bars is locked to the sync pulses and will produce a series of vertical black and white stripes within each resolution band.

Another feature of test card generators is the provision of a border consisting of alternate black and white blocks around the border of the picture area. This pattern is useful in checking correct operation of the sync circuits in the system under test and will also show up smearing effects due to poor low frequency response in the video circuits. The circle pattern if provided is very useful for checking scan linearity since it is much easier to see the non linearity as an elongation of the circle either vertically or horizontally.

6 Measuring frequency

In electronics, one widely used measurement is that of the frequency of a signal. There are a number of ways in which frequency can be measured which include various analogue techniques, but today the most popular method is the use of a digital frequency meter. The main advantages of a digital frequency meter are that it is simple to use and provides a direct readout of frequency with a precision which is better than that achieved by analogue techniques. We shall start by looking at a number of analogue methods of measuring frequency.

Using an oscilloscope

We have already seen in Chapter 4 that it is possible to measure the period for one cycle of a waveform by using an oscilloscope. Once the cycle period is known it is a simple matter to find the frequency of the signal by calculating the number of cycle periods that would occur in one second.

The procedure for frequency measurement using an oscilloscope is to set up the timebase to display one or two complete cycles of the waveform across the screen. The X SHIFT control is then adjusted so that one of the points where the wave crosses the horizontal axis of the screen graticule is aligned with the left hand edge of the graticule. The number of graticule divisions is then measured to the corresponding point on the next cycle of the wave. This figure is multiplied by the time/division setting of the timebase to give the period of the waveform being displayed. The frequency of the wave is then calculated by taking the reciprocal of the cycle period.

For example, if the timebase is set for 10μs per division and one cycle of the displayed waveform occupies 5 divisions then the wave period is 50μs. The frequency is given by

$$f = \frac{10^6}{50} = 20{,}000 \text{ Hz}$$

This method is not very precise since the average oscilloscope graticule is calibrated with only 50 divisions along its X axis. Thus if a complete cycle takes up most of the width of the graticule, the frequency obtained is probably within about 2% of the actual frequency. If it is not possible to select a range where one cycle is spread over most of the screen width, the range should be selected to give two cycles of the wave across the screen. The period measurement is then made over two cycles of the displayed wave and the result is divided by two to get the period of one cycle of the waveform.

If a high frequency is being displayed where the timebase is running at maximum speed but there are still several cycles across the screen, then the X expansion facility can be brought into play. With X expansion selected it should be possible to spread one or two cycles of the waveform out across the screen and a measurement can then be made of the cycle period. In this case the period obtained from the number of divisions per cycle will need to be divided by the X expansion factor to give the correct result.

Using a Lissajous display

In Chapter 4 we saw that by applying one signal to the X input and the other to the Y input of an oscilloscope a Lissajous figure could be produced and the phase relationship of the two signals could be found. It is also possible to use the Lissajous type display to compare the frequencies of two signals, particularly if one signal is a harmonic of the other.

If the signal on the X input has twice the frequency of that on the Y input, the result will be a figure of eight shape with two loops one above the other as shown in Figure 6.2. The actual shape will change according to the phase difference between the two signals and can become a curved vertical line at one extreme. If the Y input has twice the frequency of the X input the figure of eight will be lying on its side with the two loops alongside one another as shown in Figure 6.1. Increasing the frequency ratio between the

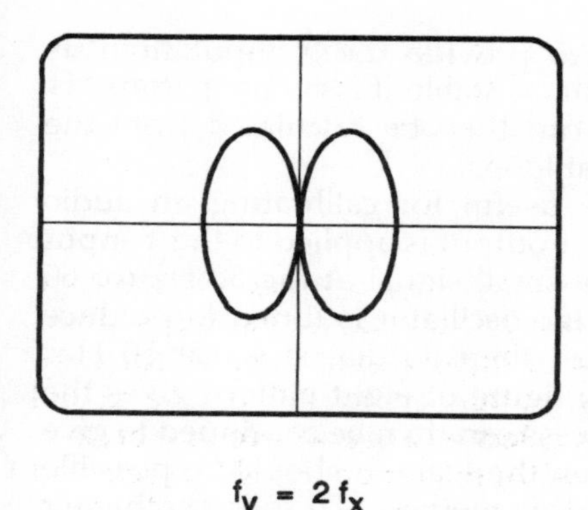

Fig 6.1 Lissajous figure for $f_Y = 2f_X$.

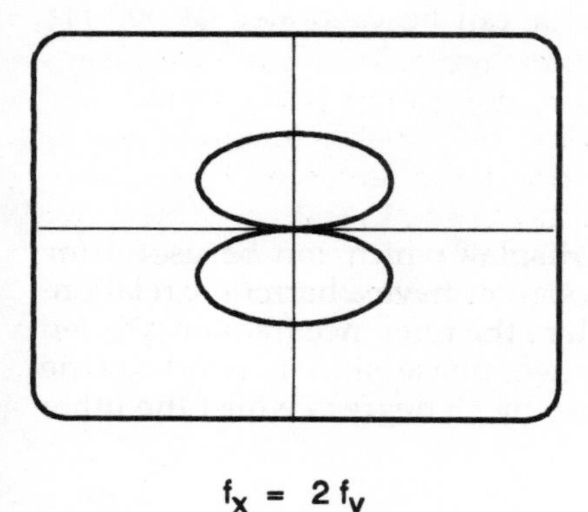

Fig 6.2 Lissajous figure for $f_X = 2f_Y$.

signals to 3:1 produces three loops in the direction of the higher frequency signal.

To use this technique to measure frequency, the signal to be measured is applied to the Y input and a known frequency is applied to the X input. The frequency of the Y input can be found by multiplying the X input frequency by the number of vertical loops in the pattern and dividing by the number of horizontal loops. This will work when the Y input frequency is either a multiple or sub-multiple of the X input frequency. Other frequencies do not usually produce a stable display pattern.

If a signal generator is used to provide the X input then its frequency can be adjusted until a stable Lissajous pattern is produced and the Y frequency can then be calculated from the number of vertical and horizontal loops.

The Lissajous display can be useful for calibrating an audio oscillator. In this case the oscillator output is applied to the Y input and the X input is driven from a small signal at the 50 Hz (or 60 Hz) power supply frequency. If the oscillator is tuned to produce a single circular Lissajous pattern then its dial is set at 50 Hz. Adjusting the oscillator to give a figure of eight pattern gives the 100 Hz calibration point. The process can then be continued to give 150, 200, 250 and 300 Hz points on the scale. It should be possible to go up to about 500 Hz using this method. To calibrate higher frequencies a simple oscillator could be built and set to say 500 Hz using the Lissajous technique with 50 Hz mains as the X input. This new 500 Hz oscillator can now be used as the X input and the dial of the variable audio oscillator can be calibrated at 500 Hz intervals up to perhaps 5000 Hz.

Broken ring display

A variant of the Lissajous type display which can be useful for frequency comparison where the signals have a harmonic relationship is the broken ring display. Here the reference frequency is fed to the X and Y plates through two phase shift networks. One network advances the signal phase by 45 degrees whilst the other

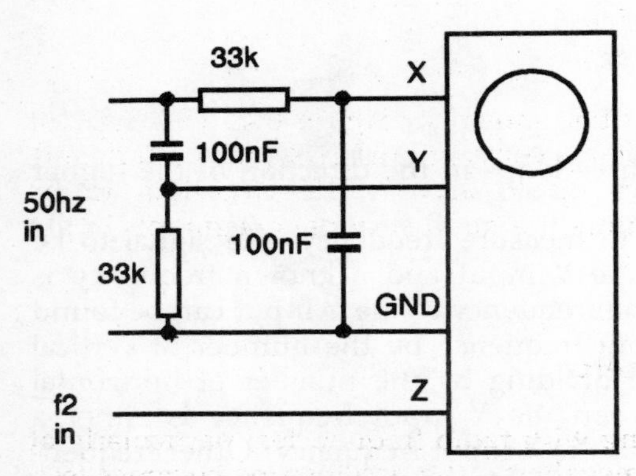

Fig 6.3 Connections for producing broken ring display.

produces a 45 degree phase lag. The result is that the two signals have 90 degrees phase difference and should produce a more or less circular display. The signal to be measured is now fed to the Z input of the oscilloscope where it will modulate the brightness of the trace. Figure 6.3 shows the connections to the oscilloscope and typical values required for producing a broken ring display using a 50Hz sine wave to generate the circular trace.

If the Z input signal has the same frequency as the X-Y drive signal then one segment of the ring will be blanked out. The darkened segment may move slowly around the ring as the two signals drift in phase relative to one another. When the Z input frequency is doubled there will be two dark segments in the displayed ring and as the Z frequency is increased further, the number of dark segments increases. When the Z frequency is not a harmonic the dark and light sectors will rotate around the ring but as each harmonic is reached the pattern will become stable to

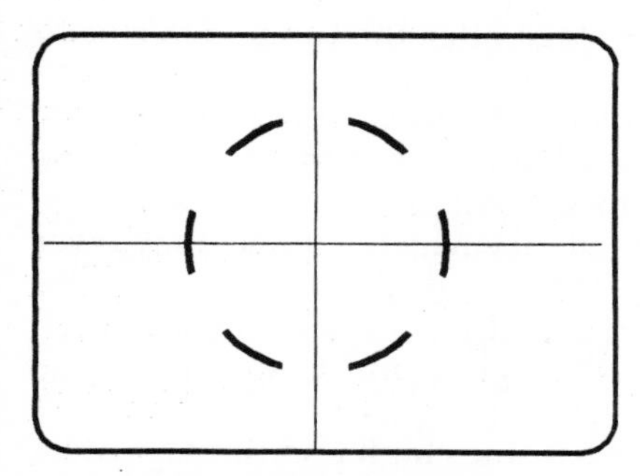

Fig 6.4 Typical broken ring display.

give a display similar to that shown in Figure 6.4. This type of display is less confusing than a conventional Lissajous pattern and should enable harmonics up to around the twentieth to be measured by simply counting the number of dark segments in the ring.

The absorption wavemeter

When we come to working with radio frequencies, particularly at frequencies above a few megahertz, the oscilloscope becomes less

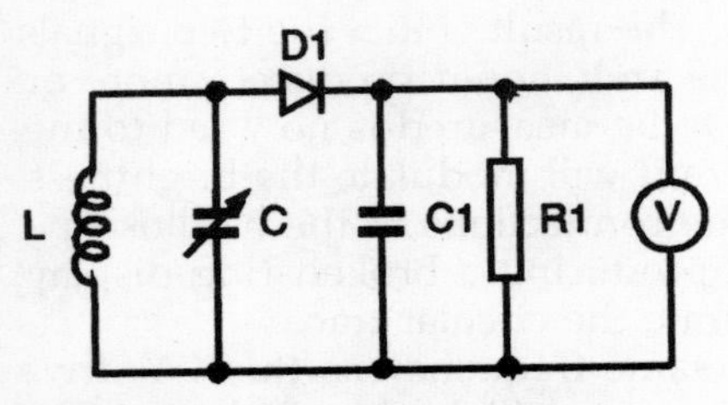

Fig 6.5 Absorption wavemeter circuit.

useful and an alternative technique using an instrument called an absorption wavemeter is often used.

The absorption wavemeter is really little more than a crystal set receiver with a circuit similar to that shown in Figure 6.5. The wavemeter circuit consists basically of a tuned circuit connected to a simple diode detector with a sensitive moving coil meter as an indicator. To achieve better sensitivity, some absorption wavemeters use a transistor amplifier stage after the diode detector to provide the current to drive the meter.

If the inductor L is loosely coupled to a source of radio frequency signals, such as an oscillator, a small amount of the signal will be injected into the tuned circuit which consists of L and C in series. When the injected signal has a frequency which is somewhat lower than the resonant frequency of the tuned circuit LC, then the inductive reactance is low but the capacitive reactance will be high and very little current will flow through L and C so the voltage across C is virtually zero. If the signal is higher than the resonant frequency the inductive reactance becomes high and the capacitive reactance low, which again gives a low circulating current and virtually no voltage across the capacitor. When the injected signal is at the resonant frequency of LC the inductive and capacitive reactances become equal but of opposite sign so they cancel out leaving only a very small resistance in the circuit. The current now rises to a maximum and a large voltage is developed across the capacitor. This voltage is rectified by the diode and gives a reading on the meter.

If the wavemeter is coupled to the signal source and the wavemeter capacitor is varied, the meter reading should rise to a maximum as the tuning of the LC circuit passes through the frequency of the signal being measured. The response should be similar to that shown in Figure 6.6 with a sharp peak as the wavemeter is tuned through the signal frequency. The best results

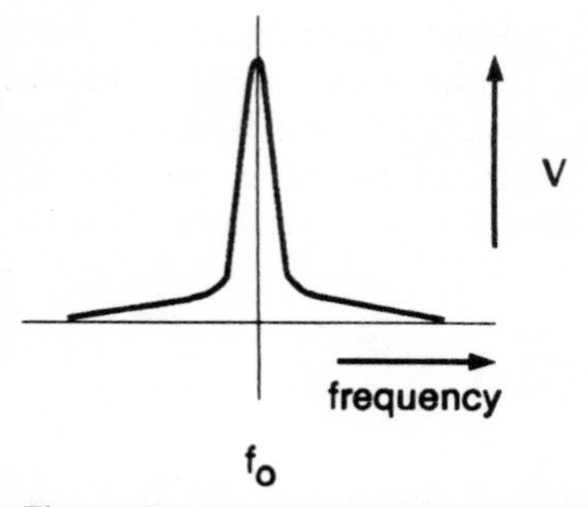

Fig 6.6 Frequency response of absorption wavemeter.

are obtained with the wavemeter only very loosely coupled to the signal under test. Usually the inductor is mounted on the outside of the case of the wavemeter and coupling is achieved by simply bringing the inductor close to the circuit which is generating the signal, or close to wires that are carrying the signal.

In a practical instrument the capacitor is fitted with a dial whch is calibrated directly in frequency. The average tuning capacitor has a ratio of about 9:1 between its maximum and minimum capacitance values and since the resonant frequency of the LC circuit is proportional to the square root of capacitance this produces a change of 3:1 in resonant frequency over the capacitor's tuning range. To provide a wider frequency coverage, the instrument normally uses plug-in inductors and the value of the inductor fitted is changed to give a series of frequency ranges which might extend from about 100kHz up to perhaps 200MHz. Typically this range of frequencies would be covered by a set of 7 or 8 plug-in inductors.

A typical absorption wavemeter is able to give only an approximate frequency reading with an accuracy of perhaps 5%. It should be noted that the absorption wavemeter will also respond to harmonics of the basic signal frequency. These harmonics will produce a smaller signal reading on the meter but this facility can be useful for checking the level of harmonic output from a radio transmitter.

An absorption wavemeter can also be used as a simple form of field strength meter by connecting a short antenna rod to one end of the tuned circuit. When the unit is used in the vicinity of a radio transmitter it will produce an indication of relative field strength if it is tuned to the transmitter frequency. This facility can be useful in assessing the radiation pattern of a beam antenna. In this case the meter is set up at a distance from the antenna and tuned for maximum output when the antenna is directed towards the meter. If the antenna is now rotated, readings can be taken at say 10 degree intervals and these can be plotted to indicate the radiation pattern of the antenna.

The dip meter

Another useful instrument for making measurements at radio frequencies is the dip meter which in some ways can be considered as the converse of an absorption wavemeter. In the absorption wavemeter some of the signal from the circuit under test is absorbed by the tuned circuit of the wavemeter and gives a peak indication on the detector meter. In the dip meter the signal is generated by an oscillator in the instrument and this signal is coupled to the circuit under test. When the oscillator is tuned through the resonant frequency of the circuit under test, some of its energy is absorbed and there is a dip in the oscillation amplitude which can be detected and displayed on a meter. This instrument is primarily used for determining the resonant frequency of a tuned circuit or an antenna. The dip meter will not work with a non resonant circuit or a circuit with a very low Q.

The block diagram of a dip meter is shown in Figure 6.7. The heart of the instrument is an oscillator, such as the Hartley or Colpitts, which is tuned by an LC circuit and may use either a valve or transistor to provide the amplification needed for oscillation.

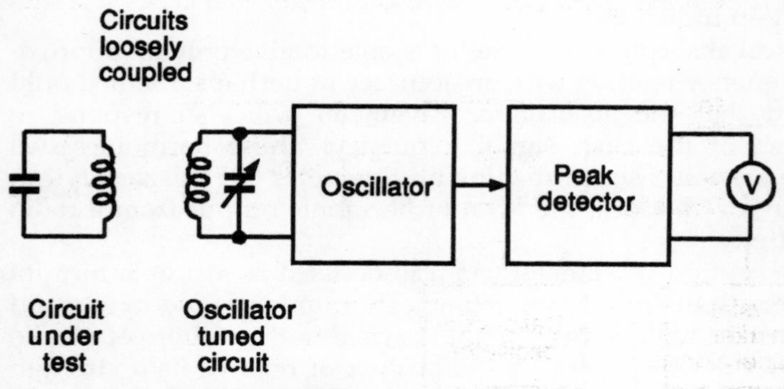

Fig 6.7 Block diagram of dip meter.

The inductor of the tuned circuit is mounted on the outside of the instrument case in the same way as for an absorption wavemeter, and the capacitor is fitted with a dial calibrated in frequency. The oscillation amplitude is detected by a diode detector and amplified to provide enough current to drive a meter. If a valve oscillator is used, grid current will flow when the circuit oscillates and the level of grid current is a measure of oscillation amplitude. In such an instrument the meter is connected directly in the grid circuit to indicate grid current and the unit is referred to as a grid dip oscillator or GDO.

To use the dip meter the inductor on the instrument is coupled to the circuit under test by placing it in close proximity so that magnetic coupling occurs. The dip meter is then tuned and when its oscillator frequency is equal to the resonant frequency of the circuit under test some energy will be absorbed from the oscillator circuit. The effect is to reduce the oscillation amplitude and hence there is a dip in the meter reading.

The dip meter can also be used as a simple frequency source since it produces a signal at the frequency indicated on its capacitor dial. If this signal were coupled to the antenna input of a receiver then the receiver could be calibrated by using the dip meter as a signal generator.

Heterodyne wavemeter

For a more accurate analogue measurement of frequency, a heterodyne wavemeter can be used. This consists of an accurately calibrated variable frequency oscillator combined with a mixer circuit as shown in Figure 6.8. The second input to the mixer is the

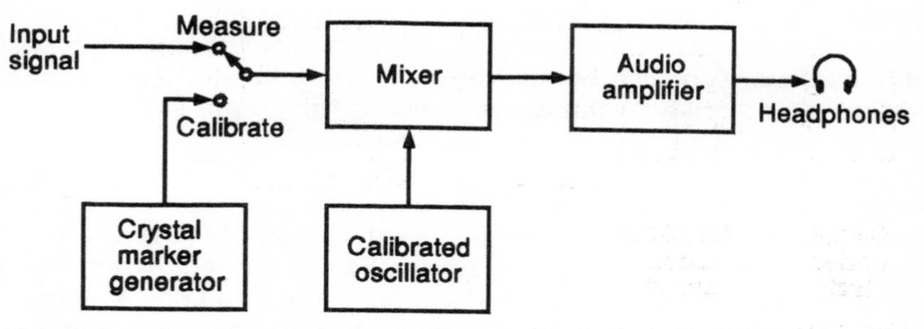

Fig 6.8 Block diagram of heterodyne wavemeter.

unknown frequency. The output of the mixer is fed to an audio amplifier and a pair of headphones.

In circuits based around valves, the mixer is usually a hexode valve in which two separate grids in turn control the level of anode current. The effect is that the anode current, and therefore the output signal, is proportional to the product of the two input signals. Similar results can be achieved by using a dual gate FET. If two sine wave signals are multiplied together in the mixer circuit, the output signal from the mixer contains four different frequency components. These four output frequencies comprise the original two input frequencies and two new signals which have frequencies equal to the sum and difference of the two input frequencies. Thus if the inputs are 100kHz and 110kHz the extra frequencies at the output will be 210kHz and 10kHz.

By adding a low pass filter at the output of the mixer all frequencies above the audio frequency range can be removed and the resultant audio frequency signal is used to drive a pair of headphones. If the wavemeter frequency is more than about 10kHz different from the frequency of the signal being measured, there is no output from the headphones. As the wavemeter oscillator frequency approaches the frequency of the input signal, the difference frequency at the mixer output will fall until it comes into the audio frequency range. This produces a high frequency audio tone in the headphones. The pitch of the tone falls as the two frequencies come closer together and reaches zero when the wavemeter oscillator is at the same frequency as the input signal. If the wavemeter frequency continues to change in the same direction the tone rises in pitch again as the difference between the two frequencies increases.

To use the heterodyne wavemeter the signal to be measured is input to the wavemeter test input and the wavemeter oscillator dial is tuned until a low pitch audio tone is heard. At this point the difference between the input signal frequency and that of the wavemeter oscillator will be a few tens of hertz. The frequency of the oscillator can then be read off from the calibrated dial of the instrument to give a value for frequency of the input signal.

The heterodyne wavemeter can also be used as a calibrating signal source in order to identify the frequency of a signal being received on a radio receiver. In this case, the output of the wavemeter oscillator is coupled to the antenna input of the receiver where it will combine with the signals being received. When the wavemeter is tuned close to the frequency of the unknown signal that is being received it will produce an audio frequency heterodyne which will fall to a low frequency growl when the wavemeter is tuned to the frequency of the incoming signal. The frequency can then be read off from the wavemeter dial. It should be noted that the heterodyne type wavemeter will also respond to harmonics of the input signal that is being measured and this must be taken into account when assessing the frequency of the signal.

One problem with the heterodyne wavemeter is that the frequency calibration of the dial may alter with time due to factors such as temperature changes. To ensure accuracy of measurement a heterodyne wavemeter usually has a built in crystal controlled calibration oscillator which will allow the dial calibration to be set accurately before making a measurement of the unknown frequency. Typically this oscillator might have crystals for 100kHz and 1MHz and its output is amplified and clipped to produce a square wave signal. This square wave signal contains a large harmonic content so that if the 1MHz crystal is switched in signals will be produced at 1MHz intervals through the range from 1MHz to perhaps 30MHz. To set up the calibration point the dial is set to one of the 1MHz harmonics and the calibrator signal is used as the input signal. A fine tuning trimmer capacitor is then adjusted to produce a zero audio frequency signal and the wavemeter is then set up ready for use. If desired, checks can be made at 100kHz intervals through the range by using the 100kHz crystal to generate the input calibration frequency.

Crystal calibrator

The heterodyne wavemeter uses a quartz crystal controlled oscillator to generate accurate calibration signals. A similar crystal controlled oscillator could also be used to provide accurate marker frequencies for checking the scale calibration of a radio receiver. An instrument of this type is called a crystal calibrator.

If a slice of quartz crystal is placed between two metal plates and a short pulse of voltage is applied across the plates, the quartz crystal will vibrate mechanically at a frequency determined by the thickness of the quartz plate. The vibration of the quartz also produces a small oscillatory voltage across the plates at the vibration frequency. If we arrange an external circuit that picks up this oscillatory voltage, amplifies it and then applies it back across the crystal in phase with the original oscillation, the circuit will oscillate continuously and produce an output signal whose

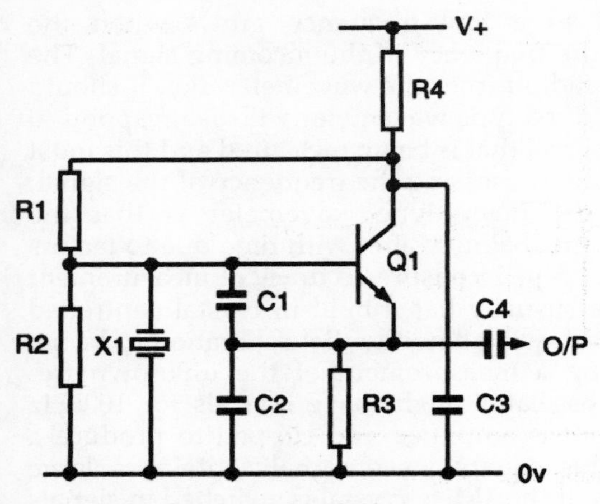

Fig 6.9 Colpitts type crystal oscillator circuit.

frequency is controlled by the dimensions of the crystal itself. In fact the crystal acts as if it were a resonant circuit with a very high Q factor and provides an accurate and stable oscillation frequency.

Figure 6.9 shows a typical circuit arrangement for a crystal controlled oscillator which, in this case, is basically a Colpitts type oscillator except that the quartz crystal (X1) replaces the conventional LC tuned circuit.

For a crystal marker generator, the quartz crystal used might conveniently be cut to oscillate at exactly 1 MHz. If the oscillator circuit used generates a reasonably clean sine wave then the main output signal will be at 1 MHz but there will probably be very low amplitude harmonic signals at 2 MHz and 3 MHz in the output as well. By clipping the oscillator signal to produce a square wave a large amount of harmonic signals can be generated with harmonics extending up to perhaps 30 or 40 MHz. Suppose this square wave signal is applied to the antenna input of a receiver. As the receiver is tuned through 1 MHz it will respond to the fundamental signal but easily identifiable signals will also be found at every other megahertz point on the receiver tuning scale going up through the bands to perhaps 30 MHz. By using a crystal controlled calibrator it is possible to check the receiver dial calibration at every exact megahertz point throughout the receiver's tuning range.

A crystal marker generator using logic gates as the amplification stages is shown in Figure 6.10. In this circuit the two logic gates

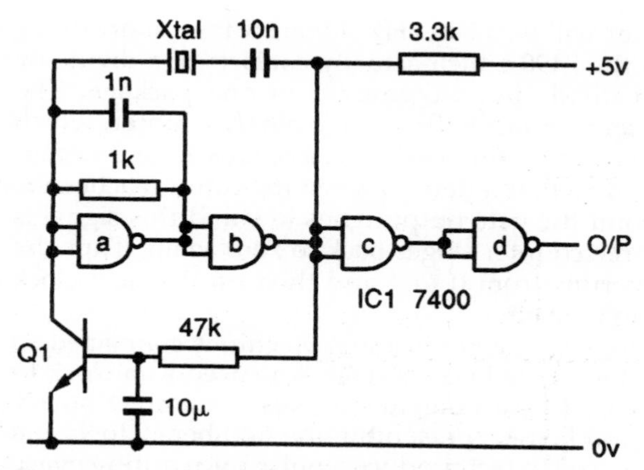

Fig 6.10 Crystal oscillator using logic gates.

IC1a and IC1b form the actual oscillator circuit whilst gate IC1c acts as a clipper and buffer to produce a square wave output. The inputs of IC1a and IC1b are biased so that they are in the indeterminate region between the 1 and 0 logic levels and the gates effectively act as a high gain amplifier. The crystal acts as a band pass filter which feeds back maximum signal at the resonant frequency of the crystal so that the circuit oscillates at that frequency. Using this circuit it should be possible to obtain useful harmonic signals of the crystal frequency up to around 50MHz.

By using a 1 MHz marker generator based on a quartz crystal we can readily find the 1 MHz divisions on the receiver scale so that the scale can be accurately calibrated at these points. The 1MHz calibration points are some distance apart and on the lower frequency ranges of a receiver this might give only one or two known frequency points on the receiver scale. It would be useful to have calibration points at closer intervals through the receiver scale. This can be achieved by having two crystals available for use in the oscillator circuit. One crystal might have a frequency of 1MHz and the second a frequency of 100kHz. By switching in the 100kHz crystal in place of the 1MHz crystal, it is possible to produce intermediate calibration points at every 100kHz along the receiver scale.

Both 1MHz and 100kHz markers can be produced using only one crystal. To achieve this, the 1 MHz square wave is fed to a digital logic counter circuit which gives a division ratio of 10:1. The output of the counter will be a 100 kHz signal. A typical divide by ten counter chip is the 7490 which actually consists of a divide by five counter and a divide by two counter in one package. The divide by five counter is in fact a three stage binary counter, which normally divides by eight, combined with a feedback gate circuit. The gate is connected so that it detects when the count reaches the value 5 at which point the gate output goes low and this signal is used to reset the three counter stages back to zero again. Thus the actual count output runs from 0 to 4 and then on the next clock pulse it goes back to 0 again.

For most applications as a counter the 7490 chip is connected so that the output of the divide by two stage A is used as a clock to drive counter stage B. This arrangement gives the correct binary outputs from ABC and D for representing the numbers 0 to 9. The output from stage D does in fact produce a pulse once during every 10 input clock periods but the waveform is at 0 for the first 8 clock periods and then goes to 1 for the last two clock periods. Whilst this signal will produce quite good harmonic outputs it would be better if the output were an equal mark-space square wave. This type of output can be achieved by re-arranging the connections of the 7490 as shown in Figure 6.11. In this case the initial clock signal is used to drive the B input of the counter and the D output is used to clock the A stage of the counter. Since the A stage is a simple divide by two circuit it will produce a square wave with equal on and off periods.

If the output of the 7490 divider circuit is now fed to the input of a receiver it will produce signals at 100 kHz intervals throughout the tuning range. This idea could be extended further by adding another 7490 divider circuit to produce 10kHz markers. If the

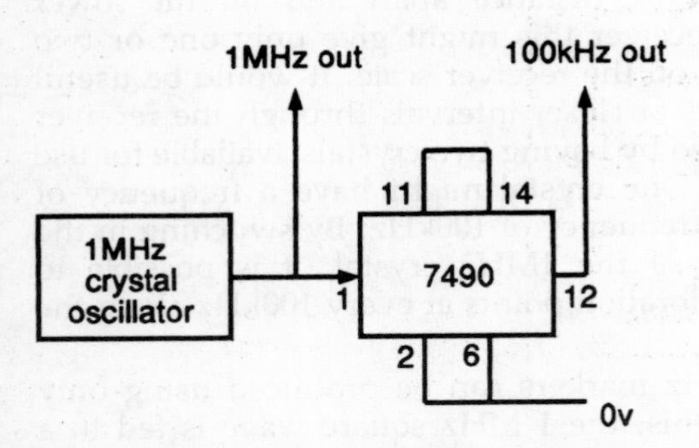

Fig 6.11 Marker generator for 1MHz and 100kHz.

receiver has a bandspread tuning dial or alternatively has an accurate logging scale on its tuning dial then the 100kHz markers are usually adequate calibration. By using the bandspread or logging dial scale it is possible to locate frequencies between the 100 kHz points to an accuracy of perhaps 5 to 10 kHz and this is sufficient to be able to identify the frequency of a broadcast signal that is being received off air.

Digital time measurement

Although it is possible to measure time periods, such as the duration of a pulse, by using an oscilloscope a more accurate method is to use digital logic techniques and to display the time period as a direct digital readout in microseconds or milliseconds.

The basic arrangement for a digital time measuring instrument is shown in Figure 6.12. The first requirement is an accurate clock generator which is usually based on a quartz crystal controlled oscillator. If the instrument is to measure time in microseconds then the clock would be set up to generate a 1 MHz square wave

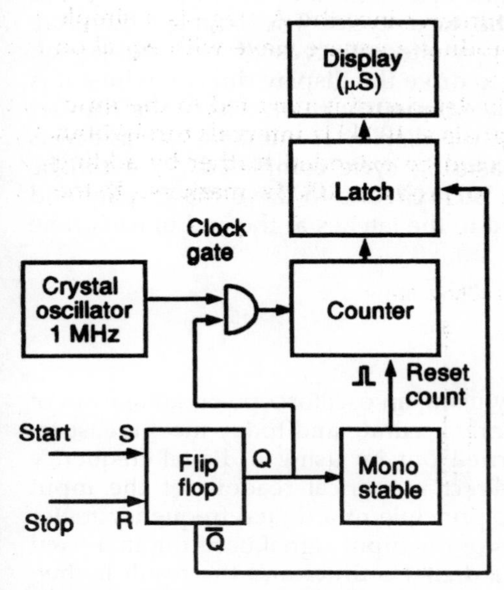

Fig 6.12 Digital period timer block diagram.

signal, and this should have an amplitude suitable for driving digital logic circuits. The process of measuring time is quite straightforward. A start pulse is generated at the start of the timing period and this triggers a flip-flop which opens a gate to allow the clock pulses through to drive a digital counter. The counter itself might typically be a four or five decade BCD counter using logic circuits. The outputs from the counter are used to drive either liquid crystal or light emitting diode seven segment displays to provide a four or five digit numerical readout. At the end of the timing period a stop pulse resets the flip-flop which closes the clock gate and stops the counter. At this point the number contained in the counter is the time in microseconds, assuming that the clock is running at 1 MHz, and this is displayed on the numerical readout. The counter needs to be reset to zero at the start of the timing period and this can be achieved by using a monostable multivibrator which is triggered by the start pulse and produces a short pulse to reset the counter before the first clock pulse arrives.

When measuring the period of a repetitive waveform such as a sine wave, the counter logic might be rearranged so that the flip-flop changes state each time an input pulse is received. Now the counter will count during one cycle period of the input wave and will be stopped during the next cycle. The counter therefore measures the duration of every alternate cycle of the input signal.

If the counter is allowed to drive the display directly whilst it is counting, the readout will continuously change until the counter stops. It would be better if the display remained static and was updated only when a measuring cycle completed. This can be achieved by using a set of latch circuits to drive the display and the count data is transferred to the latches at the end of each time measurement cycle.

Digital frequency meter

Measurements of frequency using an oscilloscope or some form of wavemeter are not particularly accurate and today most measurements of this type are carried out by using a digital frequency meter which provides a direct numerical readout of the input signal frequency. The basic principle of a digital frequency meter is that the number of cycles of the input signal occuring in a fixed time period is counted on a digital counter and the result is then displayed as a frequency reading on a digital readout. For example

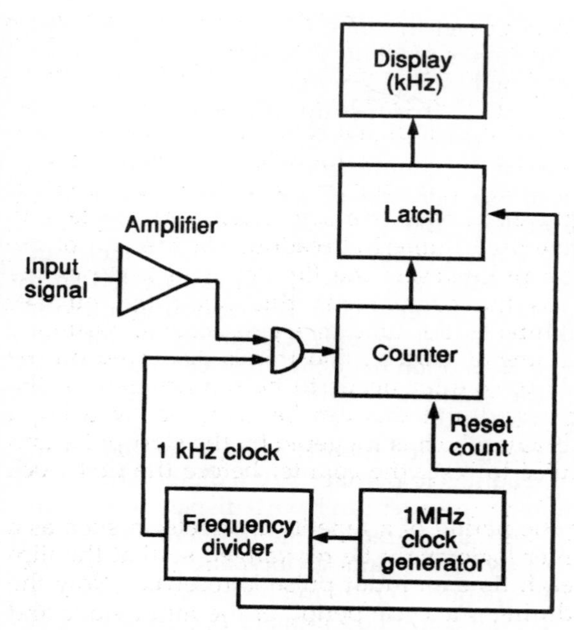

Fig 6.13 Block diagram of digital frequency meter.

if the number of cycles of input signal is counted for a period of 1 millisecond then the resultant count will be the frequency of the signal in kilohertz.

The general arrangement for a frequency meter is shown in Figure 6.13. Here an accurate clock is used to generate the gate signal so that the input signal is passed to the counter for an exact period of time. The counter will now count off the number of cycles occurring in the fixed time period and this count is used to drive a set of displays to produce the frequency readout. If the preset time period is set at 1 second then the counter will indicate a result in hertz. For audio frequency applications it might be convenient to use this arrangement and if the counter has six stages then frequencies up to 999,999 kHz could be measured. By changing the time gate to say 0.01 second, the readout could be 100 Hz steps and the 6 digits would now allow frequencies up to 99.9999 MHz to be displayed.

If the counter is built up from TTL or Low Power Schottky it should be possible to measure frequencies up to perhaps 50 MHz.

For frequencies above this level the time delays in the gates can produce errors in the frequency readout. One solution might be to build the counter using high speed emitter coupled logic (ECL) devices which can count at speeds up to several hundred megahertz. In fact it is not necessary to build the entire counter using ECL because most of the counter stages are operating quite slowly. The solution is to use a prescaler counter with a frequency division ratio of 10 or perhaps 100. If the prescaler has a division ratio of 10 and the input is a frequency of 500 MHz then the output frequency will be just 50 MHz and this signal can be fed to a standard TTL type frequency meter circuit. The readout frequency must now be multiplied by 10 to give the input frequency. Commercial frequency meters often have the prescaler built in and the decimal point on the readout is shifted so that the correct frequency is displayed.

The main advantage of the digital frequency meter is that it can provide high precision readings to perhaps 5 or 6 digits. The more expensive types of frequency counter can have a display resolution of 8 or more digits. The actual accuracy of the frequency readings depends upon the accuracy of the timing clock and the speed at which the input gate can operate. The timing oscillator is normally based on the use of a crystal controlled primary oscillator so that the clock accuracy can be high and the clock frequency stable.

Frequency meters usually have an adjustable sensitivity control which sets up the level at which the input signal will trigger the counter. This control can be adjusted so that the meter does not respond to noise or other low level interference. In laboratory instruments, the frequency measuring function is often combined with the ability to operate as an event counter or as a period timer.

Frequency offset

Frequency counter circuits which are built into receivers to give digital frequency readout are usually driven from the local oscillator signal that drives the receiver mixer stage. To obtain the correct frequency readout the basic digital frequency meter circuit has to be modified in this application.

In a typical AM receiver the local oscillator signal is offset 455 kHz higher than the input signal frequency to produce a constant difference frequency of 455 kHz at the output of the mixer stage. This fixed 455 kHz frequency is then amplified by the intermediate frequency (IF) amplifier and detected to produce an audio signal. If the local oscillator signal is used to drive the frequency meter, the display readout will be 455 kHz higher than the frequency of the station being received. To correct this, 455 must be subtracted from the frequency reading. This is usually done by presetting the counter to −455 instead of 0 at the start of each measurement cycle. The counter now has to count off 455 cycles before it reaches zero on its frequency count and this produces the same result as subtracting 455 from the count.

An FM receiver normally has an IF of 10.7 MHz so its local oscillator will be tuned 10.7 MHz higher than the frequency of the received station. If the frequency meter is set with a gate time of 100 μs the count will be in tenths of a megahertz so 107 will have to be subtracted from the counter value to give the correct frequency readout. A television receiver uses an oscillator offset of 39.5 MHz in Europe or 45 MHz in North America.

General purpose digital frequency meters often have a facility for presetting an offset frequency which is subtracted from the digital readout to give a correct reading of received frequency when the meter is fed from the receiver local oscillator.

7 Other equipment

Resistance boxes

Where experimental work is being carried out a switched resistance box can be a useful accessory. The ideal arrangement is a true decade resistance box giving perhaps three decades of selectable resistance. The basic circuit arrangement of this type of resistance box is as shown in Figure 7.1. For simplicity, the diagram shows only two decades. In this arrangement the box provides a range of resistance from 0 to 9.9kΩ in 100Ω steps. A typical box might have four banks with the lowest giving 10Ω steps and the highest giving

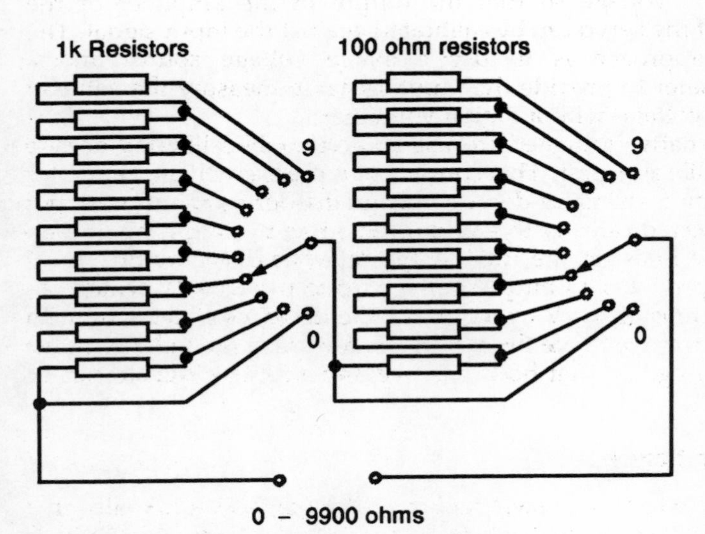

Fig 7.1 Arrangement of a decade resistance box.

10kΩ steps which would allow resistance values from 0 to 99.99kΩ to be selected in 10Ω steps.

Thus in the 10k bank each resistor has the value 10kΩ. At the 0 position, the bank is shorted out but as the switch rotor is moved 10k, resistors are added in series between the rotor and the input terminal. The output of the 10k bank switch feeds the top end of the 1k resistor bank and here the switch adds a selected number of 1k resistors in series. The 100Ω and 10Ω banks are wired in the same way and finally the wiper of the 10Ω selector switch comes out to the other input terminal of the resistance box.

The switches can be decimal type thumbweel switches and the resistors in this type of box need to be at least 1% tolerance metal oxide types to give useful results. For a home constructed unit using 1% components only the two most significant digits of the reading on the selector switches should be considered valid when assessing the value of resistance. In a commercial resistance box, the resistors are usually 1% tolerance components which have been measured and selected to give the correct values to within 0.1% or better.

DC voltage calibrators

If experimental work is being carried out on, say a dc amplifier or a servo system, it is useful to be able to input a series of known steps of dc voltage so that the output of the amplifier or the position of the servo can be calibrated against the input signal. The simplest approach is to use a stable voltage source and a potentiometer to provide the input signal to measure the value of the input voltage with a digital voltmeter.

An alternative scheme is to use an accurately calibrated decade voltage calibrator unit. This consists of a precise voltage generator followed by a switched decade voltage divider so that the output can be selected directly to any desired value by simply setting up the decade dials on the unit. A typical voltage calibrator unit of this type provides an output voltage up to perhaps 2V which can be set in steps of 1 mV by adjusting the decade switches. In most cases the unit will have the facility for reversing the polarity of the output voltage so that both positive and negative signals can be generated.

Capacitor boxes

It is possible to have a switched capacitor box which operates in a similar way to the resistor box. In this case the capacitors in each decade are connected successively in parallel to produce the desired value of capacitor and the total capacitance of each decade is connected in parallel with that of the other decades. Because of stray capacitance effects, about the lowest increment of capacitance that is practical would be 100pF. Thus a box could be built with the first decade going up to 1nF and successive decades to 10nF, 100nF and 1μF respectively. For the lower decades polystyrene or silver mica capacitors with 2% tolerance might be used to give reasonable accuracy and good stability. For the higher ranges metallised polyester film capacitors of 5% tolerance might be used.

Testing diodes and rectifiers

The basic function of a signal diode or rectifier and its polarity can be checked quite easily by using an analogue multimeter switched to its ohmmeter mode. If the black (−) lead of the meter is connected to the anode of the diode and the red (+) lead to the cathode a low resistance reading should be obtained. It may seem odd that the negative lead should be connected to the diode anode in order to make the diode conduct current. The reason for this is that most analogue meters output a positive voltage on the black (−) test lead so that when the leads are connected in this fashion the anode of the diode will be positive with respect to the cathode and the diode will conduct giving a low resistance reading on the meter. For typical germanium diodes the resistance reading will usually be about 500Ω and for silicon diodes the reading is about 2000Ω. This effect is mainly due to the higher forward voltage needed to make a silicon diode conduct. Actual values obtained will depend upon the particular type of multimeter used.

If the meter connection to the diode is reversed the diode will be reverse biased and should give a very high resistance reading. For a germanium signal diode values from about 200kΩ up to several megohms will be obtained. Low resistance values indicate a high leakage current. For a silicon diode or rectifier the reverse resistance should be several megohms. Zener diodes may indicate a low reverse resistance if the ohmmeter uses a high voltage battery as its energising source and the potential developed across the diode is higher than its Zener voltage.

To find the polarity of an unmarked diode, the ohmmeter is connected across the diode and if the resistance reading is high, the meter leads are reversed. Once a low reading has been obtained, the anode terminal will be the one connected to the black

(−) lead of the meter. A faulty diode that has gone open circuit will give a very high resistance for both polarities of ohmmeter connection. In some cases where the internal junction is fused, a faulty diode will show a low resistance or even a short circuit in both directions.

Some digital multimeters, when operated as ohmmeters, may not produce good results on these simple diode tests. This is because the voltage generated by the meter is not sufficient to make the diode conduct so that high resistance readings may be obtained for both polarities of connection of the diode. A further point to note is that many digital voltmeters output a positive voltage on the red (+) test lead.

Most of the currently available models of digital meter include separate facilities for testing diodes and transistors. In this case the diode is connected across a separate set of test terminals and the meter range switch is set to the diode test position. This test will allow a basic diode polarity check and will identify open circuit or short circuit devices.

Another test which may be useful on a diode or rectifier is a measurement of the reverse breakdown voltage. This can be checked by using the simple test circuit shown in Figure 7.2. Here

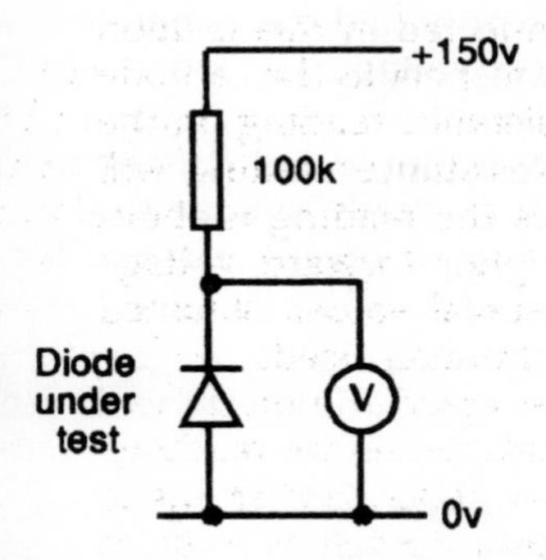

Fig 7.2 Test circuit for diode reverse breakdown voltage.

a high reverse voltage is applied to the diode through a current limiting resistor. A voltmeter connected across the diode indicates the breakdown voltage. The voltmeter should be a high impedance type (20k/V) and for a typical rectifier the resistor is arranged to limit the current to about 1mA. When the diode is connected into circuit with the anode connected to the zero line, the voltage applied will cause reverse breakdown in the diode and it will operate in much the same way as a Zener diode. The voltmeter will indicate the reverse breakdown voltage of the diode under test. For a typical rectifier diode, a test voltage of about 150V might be used since most commonly used diodes have reverse break-down voltages of 50V or 100V. For high voltage types such as the 1N4003 to 1N4005 a test voltage of up to 1000V would be required and some care is needed when handling tests with these high voltages.

A bridge rectifier can be checked both for function and reverse breakdown by testing each of the four diodes in the bridge unit as a separate diode. The connections for an unmarked bridge rectifier unit can readily be worked out once the diode polarities have been checked. The connection with two diode cathodes joined to it is the + output lead and the one with two anodes connected to it is the − output lead. The other two connections to the bridge are for the AC input.

Tests on Zener diodes

A Zener diode will normally produce the same results on an ohmmeter as a conventional silicon diode provided that when it is connected in the reverse direction the test voltage applied is not higher than the rated Zener voltage for the diode.

To check the Zener voltage of diode a simple test circuit as shown in Figure 7.3 can be used. This is basically similar to the reverse breakdown circuit except that a higher reverse current is

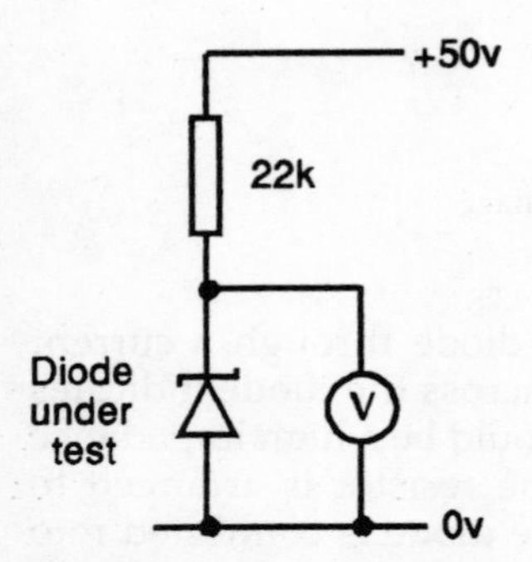

Fig 7.3 Circuit for checking Zener diodes.

allowed. The current limiting resistor can conveniently be set to limit the current to perhaps 2 mA for an approximate check on the diode Zener voltage.

The actual value of the Zener voltage will in fact vary slightly with the current drawn by the diode. If an accurate test of Zener voltage under some specified working conditions is required then the series resistor can be adjusted to produce the specified current in the diode. This can be done by making part of the series resistor variable and including a milliammeter in series to measure the current flowing into the diode.

Basic tests on transistors

A simple analogue ohmmeter can be used to check the pin connections of a junction transistor and to find out whether it is an npn or pnp type. It may also be possible to identify whether an unmarked or unknown type is a silicon or germanium transistor. Simple checks can also indicate a faulty transistor where one or both of its junctions are either short circuited or open circuit.

For lead identification and npn/pnp polarity checks the transistor can be considered as if it were a pair of diodes connected back to back as shown in Figure 7.4. In an npn type transistor, the diodes will conduct when the base terminal is made positive relative to either the emitter or collector terminals. To identify the connections connect the black (−) lead of the ohmmeter to any one of the

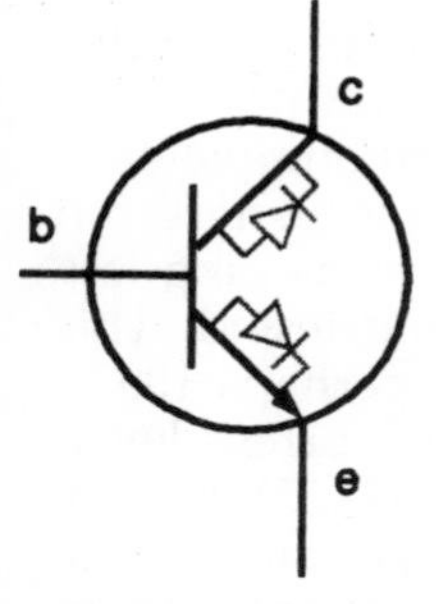

Fig 7.4 Junction transistor considered as two diodes.

three transistor leads and test the resistance reading to the other two leads. If both connections indicate high resistance, move the black lead to another transistor lead and repeat the test. If this also gives high resistance, repeat the test using the third lead of the transistor. When the resistance readings are both low the negative test lead is connected to the base terminal of the transistor. If all three tests give high readings, the transistor is either faulty or it is a pnp type. To test a pnp type the same sequence of tests is carried out but this time the test meter leads are reversed. Once the base lead has been identified it is usually possible to identify which of the other leads are collector and emitter by looking up typical lead layouts for transistors in the same type of device package or with similar lead layout to the type under test.

A measurement of resistance between emitter and collector terminals should indicate a high resistance with the ohmmeter leads connected either way round. The testing process if to make measurements of resistance between pairs of terminals with the meter leads connected one way round and with the leads reversed. The base – emitter and base – collector connections should give a low resistance for one polarity of the meter and a high resistance for the reverse polarity.

The base terminal will be the one which has a low resistance to the other two terminals in one direction, and the polarity of the base terminal in the low resistance condition will indicate whether the transistor is of pnp or npn type. If the base terminal is positive; the transistor is an npn type whilst if the base is negative; the transistor is a pnp type.

Generally for silicon type transistors the resistance between emitter and collector will be extremely high (>1M) with either polarity of the meter leads. Germanium type transistors may indicate much lower values of resistance due to their leakage current and typical values might range from around 100k upwards. Germanium transistors also tend to show lower values of reverse resistance on the be and bc tests.

Some precautions may be required when using an ohmmeter to test diodes and other semiconductor devices. Some meters can produce currents of the order 10 – 20 mA when measuring low resistances and this could damage the device being tested. This problem can be overcome by adding a resistor of perhaps 100 – 150Ω in series with the meter test lead to limit the current level. Another point to note is that some meters incorporate a 22V battery for some resistance measurement ranges and this could cause problems with transistors which have a low value of breakdown voltage.

Transistor h_{fe} testers

Although an ohmmeter test can identify connections and show up transistors which are total failures, a more useful check on the function of a transistor is to measure its h_{fe} characteristics under dc operating conditions.

The parameter h_{fe} is a measure of the ratio of the change in collector current to change in base current when the transistor is connected in the common emitter configuration. A simple test rig for testing h_{fe} is shown in Figure 7.5.

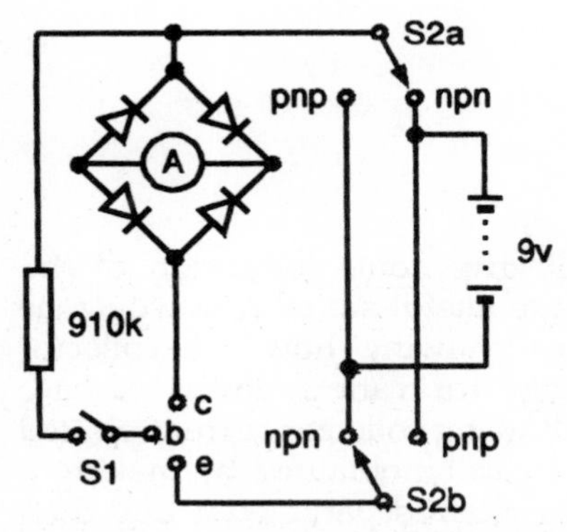

Fig 7.5 Test rig for measurement of h_{fe} of npn transistor.

The switch S2 should first be set according to the type of transistor being tested, since npn and pnp types are designed to operate from opposite polarity supply voltages. In the case of an unmarked transistor a simple test with an ohmmeter can determine if it is a pnp or npn type.

With switch S1 open circuit the collector emitter leakage current I_{ce0} can be checked. For a silicon transistor this will usually be zero but for a germanium transistor some current will flow depending upon the ambient temperature. Even with a germanium type the current should be small if the transistor is to be of any practical use.

When S1 is closed a current of 10μA is fed into the base and this will cause a collector current to flow. The collector current will be approximately equal to the base current multiplied by the h_{fe} of the transistor. Typical general purpose transistors might be expected to have an h_{fe} of 50 to 500. If the meter is rated at 5mA then the value of h_{fe} can be read off the meter scale with a full scale deflection indicating an h_{fe} of 500.

Many of the digital type multimeters that are commercially available include facilities for testing the h_{fe} of npn and pnp transistors under dc conditions. A set of three sockets is provided into which the leads of the transistor under test are inserted and the selector switch is then set to the h_{fe} test position. On some meters, separate sockets are provided for npn and pnp type devices and the transistor under test is placed in the appropriate socket. Other types use a common set of test sockets for all types of transistor and the connections are switched internally by the range selector switch which will have positions for both npn and pnp type devices.

Transistor curve tracers

Although a simple test of h_{fe} will give some indication of the characteristics of a transistor, a more useful set of data would be the characteristic operating curves showing how the collector current varies with collector voltage for various levels of base current or alternatively the graph showing collector current plotted against base current. These curves can be obtained by making a series of static measurements of the desired parameters and then plotting the results on paper, but a more convenient arrangement would be to have the measurements made automatically and the curves displayed on say a cathode ray tube screen.

To plot the curve of collector current against collector voltage at various levels of base current, a scheme similar to that shown in Figure 7.6 could be used. This produces a display of the transistor's collector current against collector voltage curves at a number of different base currents as shown in Figure 7.7. Here a full wave rectified sine wave is applied to the collector of the transistor under test and is also applied to the X input of the oscilloscope to provide a horizontal sweep signal. The horizontal axis of the oscilloscope now represents collector voltage and the scaling of the graticule can be adjusted by altering the X gain to give the desired calibration in V/division. The collector current is measured by inserting a low value resistor in series with the ground tap of the full wave transformer that provides the sweep signal. The voltage across this resistor is proportional to collector current and this signal is used to drive the Y amplifier. Again the sensitivity of the

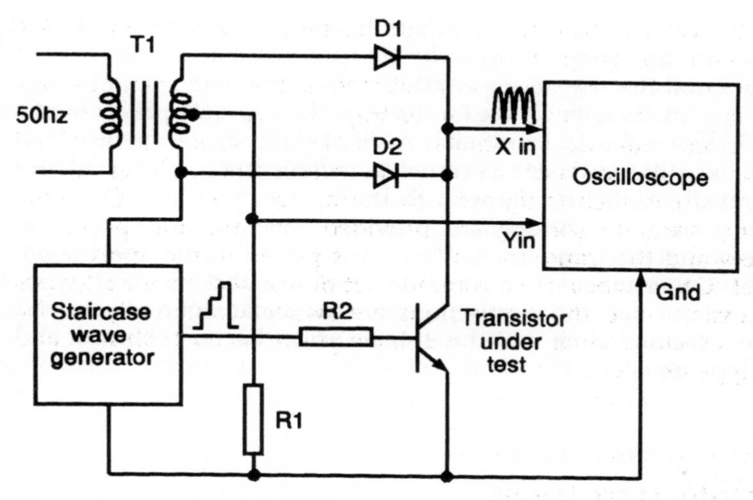

Fig 7.6 Block diagram of curve tracer set up.

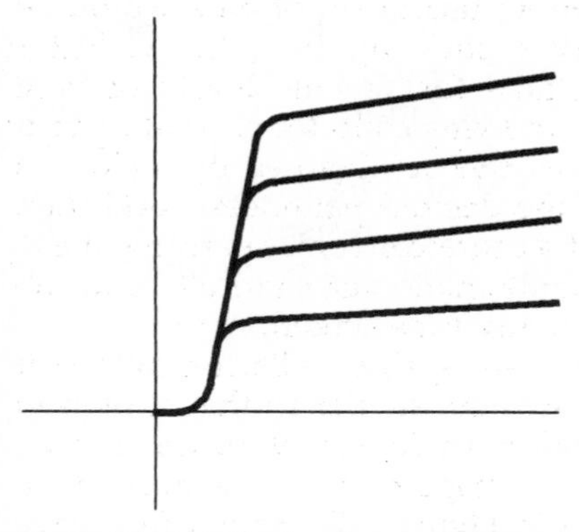

Fig 7.7 Typical display from a curve tracer.

Y amplifier is adjusted to give the desired calibration in mA/ division. For a single curve, the base current is fed via a high value resistor from a voltage source and the current is set to a desired level. For a family of curves, the base current is derived from a staircase voltage waveform which produces an increase in base current of say 1mA for each step. The steps are triggered by the sweep signal being applied to the X input so that after each scan the base current rises by 1mA so that on the next sweep a new characteristic curve is drawn. The staircase must also be reset after the required number of traces has been drawn and then a further set of curves is drawn. If the sweep is derived from the 50Hz mains supply then it should be possible to display five or six characteristic curves at different base currents simultaneously on the screen without too much trouble with flicker.

One advantage of the dynamic curve plotter over static measurements is that it becomes possible to plot points beyond the normal dissipation limits of the transistor because the overvoltage or overcurrent condition is only present for a fraction of a second and does not allow time for the transistor to become heated and possibly damaged as it would be when static readings are taken.

Logic level probe testers

In modern logic systems, the signals in the circuits are normally switched between two defined states which are usually referred to as the 0 and the 1 states. For a typical logic system, the 0 state will have an almost zero voltage level whilst the 1 state may be perhaps +4V to +5V. Although a meter could be used to check the logic state at various points around the circuit a more convenient device is a logic level probe indicator.

This instrument can consist simply of a light emitting diode connected in series with a current limiting resistor as shown in Figure 7.8. The ground return lead is connected to ground on the

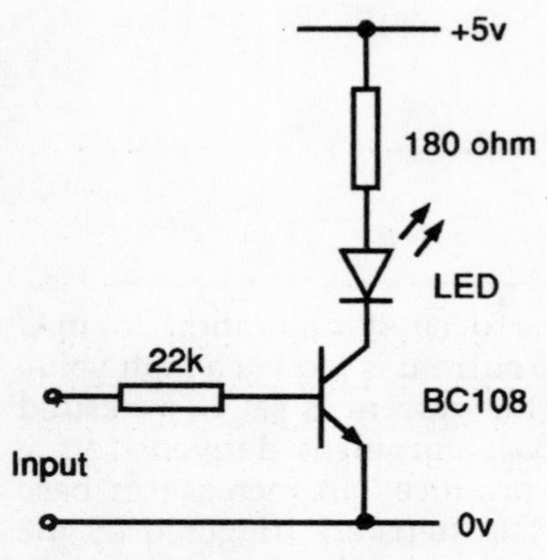

Fig 7.8 Simple logic probe.

circuit being examined whilst the top end of the resistor is connected to the test prod on the probe unit. When the prod is touched to a point in the logic circuit the logic state is indicated by the state of the led. When the logic level is 0 the diode remains off but when a logic 1 state is detected the diode will light up. To avoid loading of the logic system under test, the logic probe could use a 74LS series TTL gate or a CMOS gate as a buffer between the input and the light emitting diode.

The logic probe will also give some indication if the circuit is switching from one state to the other. If the switching rate is slow, perhaps up to about 2 or three pulses per second, the diode will flicker in sympathy with the signal. If the rate of change is higher, the diode will tend to produce a lower brightness than it would for the normal 1 state signal. In fact the brightness will tend to be proportional to the amount of time that the signal is at the 1 state relative to the amount of time it is at the 0 state.

An extension of the simple logic probe is a logic clip device which can be clipped on to an integrated circuit so that it picks up the signal levels from all of the pins of the circuit package and displays the logic state of each pin on a separate LED.

Logic analysers

The ultimate tool for diagnosing problems in logic systems is the logic analyser. This instrument provides facilities for sampling the logic states of a large number of separate input lines and displaying the information on a cathode ray tube display. Figure 7.9 shows the basic block schematic for a logic analyser. A typical instrument might have 16 to 64 input lines and will take perhaps 1000

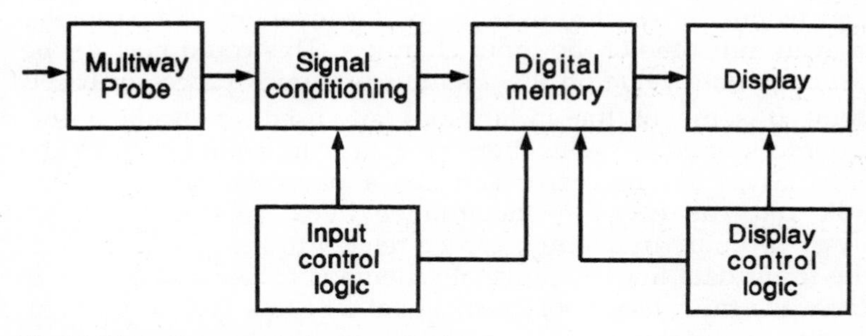

Fig 7.9 Block diagram of a logic analyser.

successive samples from each. The samples are stored in a digital memory and may then be displayed in various ways so that the operation of the system under test can be analysed. Input data lines are usually arranged in groups of eight lines and each group has a remote pod near the test clips which are connected to the circuit under test. This pod contains a set of logic buffers which are used to drive multiway cable that takes the signals back to the instrument. By buffering the signals at the remote end of the cable, any loading or capacitance effects of the cable on the circuit under test can be eliminated.

The logic analyser is almost indispensible when fault finding is to be carried out on a microprocessor based system since it enables the data bus signals to be examined in a way which is not possible with a conventional oscilloscope. The main problem with signals on a data bus is that they are multiplexed and it is rare that a repetitive pattern of data will be produced which can be examined on an oscilloscope. The other problem is that there may be 8 or 16 bus lines and these need to be examined as a group in order to make sense of the signals on the data bus. The logic analyser allows all of the data bus lines and perhaps the address bus lines to be examined simultaneously and a large sample of the signals can be stored and then be stepped through one clock period at a time to see just what is happening on the bus.

Some form of time reference is needed when the logic analyser is acquiring a block of data. A number of different techniques may be used to provide the trigger which starts the acquisition of a block of data samples. In the simplest case one channel is allocated for synchronisation and a single pulse is applied to that input when the data acquisition is to start. An alternative is where a number of input channels are examined in combination and the data acquisition starts when a particular signal combination occurs on these selected input channels. This arrangement is convenient for use where the synchronisation point is to be tied to a particular data event on some of the input channels. This could perhaps be a particular data value on the data bus of a microprocessor based system. If some of the input lines are used to examine the microprocessor address bus then the data acquisition can be made to start when the processor accesses a particular address. The sample rate can either be determined by a clock in the logic analyser or, more frequently, can be related to the timing on one of the input data lines. The usual scheme is to use one data line to carry the microprocessor clock signal and this is then used to determine when the data samples are taken.

Three forms of data display are usually provided on logic analyser systems. In the simplest form, the data for each channel is displayed as a sequence of 1 or 0 values with each digit corresponding to one sample value. For an 8 channel system the string of 1 and 0 digits for one sample across the 8 channels is displayed as an 8 digit binary number and successive samples are displayed on successive lines of the screen display. This is convenient when examining a data bus because the complete data value for each sample of the bus signal is presented as a binary number. In a system with more than 8 lines the group of data bits from each pod is displayed as an 8 bit binary number and the groups are separated on the screen.

Many logic analysers permit an alternative form of data display to be used in which the 8 bit binary sample is displayed as a hexadecimal number. This hexadecimal display mode can be particularly useful for 16 or 32 bit wide samples from say a microprocessor data bus where a string of binary bits could be rather confusing. This is shown in Figure 7.10.

Some logic analysers can be fitted with special personality modules which will convert the digital signal acquired from say a microprocessor data bus into assembly code mnemonics as well as a hexadecimal code. This makes it possible to present a disassembled listing of the instructions that are being executed by the processor. This type of listing is useful because it can be compared with the actual program that the processor is supposed to be running and faults are easily detected.

The third form of display provided on a logic analyser display presents each input as a simple oscilloscope type waveform trace which shows the logic levels and transitions against time. The general arrangement of this type of display is shown in Figure 7.11. Usually there are up to 16 separate traces displayed on the screen and these can be allocated to any desired selection of the input channel lines by programming the switches on the front panel of the instrument. This type of display is useful for examining the time relationships between events on two or more channels. Since the data acquired into the instrument's memory is perhaps 1000 samples it is usual to be able to expand the display so that a segment of the complete acquisition period is expanded to fill the width of the screen. This operates in a similar way to X expansion on an oscilloscope and effectively provides a window into the stored information. The displayed portion can then be moved through the stored data by altering the point in memory at which the display starts until the displayed section contains the events

A000	10000000
A002	11101010
A004	00101001
A006	00111011
A112	10001111
A114	10111011
A116	00000011
A118	11111001
A11A	10101010
A11C	00001000
Pod 1 signals 16 lines (Hex.)	Pod 2 signals 8 lines (binary)

Fig 7.10 Numeric display on logic analyser.

which are of interest. This action is similar to the X shift on an oscilloscope but on a logic analyser it is often controlled by a pair of push buttons which allow forward or reverse movement to be selected and releasing the button freeezes the screen display.

Another function which can be performed by many of the more expensive logic analysers is signature analysis. In this case the data patterns acquired from the equipment under test are compared with reference patterns for a piece of equipment which is working correctly. If the patterns match then the equipment under test is working normally but if they do not then the equipment has a fault.

For low frequency applications, a personal computer can be made to perform the function of a logic analyser provided that it has one or more parallel digital input ports. In this case the data words are read in from the ports at regular time intervals, determined by the computer timing clock and these are stored in the memory. When a sequence has been sampled, the data held in memory can be read and displayed either as numerical information or as a graphical display showing the waveforms for the separate logic input channels.

Spectrum analysers

One of the more complex instruments available to the electronics engineer is the spectrum analyser. This instrument uses a cathode

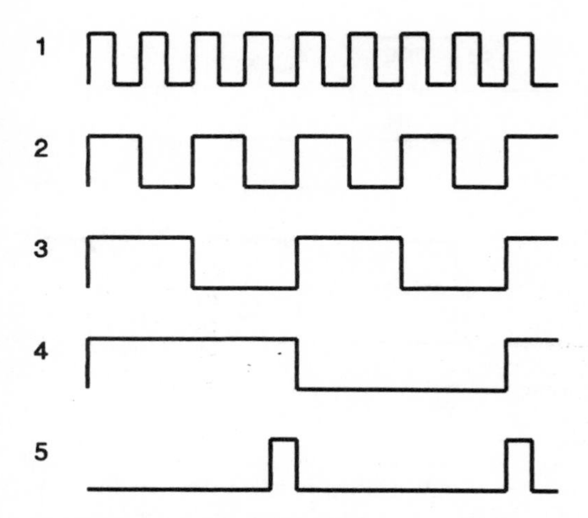

Fig 7.11 Waveform display on logic analyser.

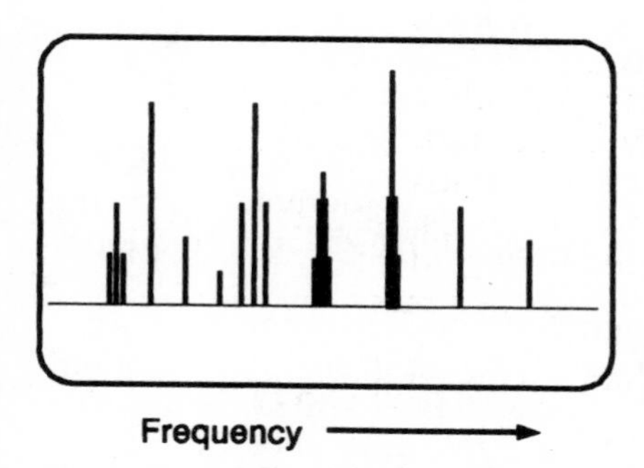

Fig 7.12 Display on a spectrum analyser.

ray tube display similar to that of an oscilloscope but instead of producing a plot of voltage amplitude against time this instrument displays voltage amplitude against a base of frequency.

If a signal is input to the spectrum analyser, the display will show the various frequency components in the signal and their relative amplitudes. Thus a pure sine wave will produce a peak output on the screen at a point along the X axis corresponding to its frequency f. If the signal also contains harmonics of the sine wave these will show up as smaller peaks on the display at their appropriate 2f and 3f positions along the X axis. A typical spectrum analyser display is shown in Figure 7.12.

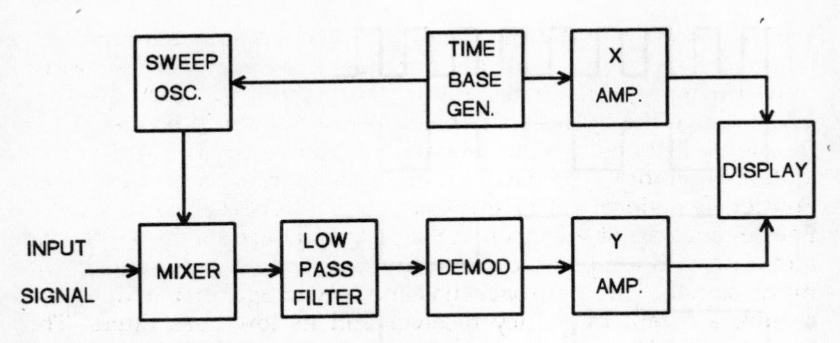

Fig 7.13 Block diagram of spectrum analyser.

The basic spectrum analyser block diagram is shown in Figure 7.13. The instrument is in effect a narrowband receiver whose tuning is swept through a range of frequencies and which produces output responses each time it comes across an input signal throughout the frequency sweep. The receiver may be a direct conversion type where the oscillator signal is mixed with the incoming signals and the output of the mixer is passed to a low pass filter and audio amplifier to produce the Y drive signal for an oscilloscope display. The X drive signal is derived from the sawtooth signal that is used to sweep the oscillator frequency through its range.

As the oscillator sweeps through its frequency range the output from the filter will indicate the level of any input signal component at that frequency. Thus if the input signal is at 1MHz then as the sweep oscillator approaches 1MHz the difference frequency component will fall in frequency until it comes inside the bandpass of the filter. At this point the amplitude of the filter output rises and will reach a peak when the sweep oscillator is at 1MHz, then the level falls off again to a level corresponding to the general broadband noise level of the input signal.

If the spectrum analyser were connected to an antenna and set to sweep through the medium wave broadcast band it would display a peak at the frequency of each of the broadcast stations that could be received at the time the test is made. The signals would have different amplitudes according to the strength of individual station signals and there would be some flutter in amplitude caused by the modulation on the signal. Spectrum analysers are useful for experiments on radio transmitters to check for harmonics or spurious radiation from the transmitter.

More modern spectrum analysers use digital techniques in which the signal is sampled at a frequency which is at least three times the highest frequency of interest and about 1000 samples are taken. Since the input is an analogue one the input has to be fed through a fast analogue to digital converter before being stored in a digital memory. The data in memory is then processed by a Fast Fourier Transform (FFT) process which effectively carries out a Fourier analysis of the incoming signals and outputs the amplitude and if desired the phase of all of the frequency components in the input signal. This approach has the advantage that it does not require a swept frequency receiver and its low pass filters. The limitation of this digital approach to spectrum analysis is the speed at which data samples can be taken and converted into digital data which limits the upper frequency of the analysis.

8 IEEE488 interface bus

Until the mid 1970s, electronic test instruments had always been designed as stand alone units capable of performing one particular function. A few manufacturers experimented with composite instruments where perhaps a signal generator and an oscilloscope were built into a single case. This idea was useful in portable sevicing equipment since it reduced the number of items of equipment that had to be carried by an engineer.

At this time, microcomputers had become widely available and were beginning to be incorporated in some of the more sophisticated laboratory instruments. With this in mind, Hewlett Packard, one of the major instrument manufacturers, came up with a new concept in instrument design. The basic idea was to have all of the instruments connected to a common bus system so that their operational functions could be controlled by commands over the bus. At the same time, instruments could also exchange data via the bus and results could be collated by a small computer to produce printouts or graphs of the results of an experiment. By using this bus system it becomes possible to automate an experiment or test procedure thus reducing the time taken and eliminating the need for manual effort in taking readings and compiling results.

The bus system was originally called the Hewlett Packard Instrument Bus (HPIB). After a few minor modifications the bus system was later adopted as a standard by the American IEEE and was named as the IEEE488 General Purpose Interface Bus (GPIB). Today most of the more advanced instruments designed for professional or laboratory use are fitted with this bus system either as a standard feature or as an optional add-on feature.

In a typical application, a system consisting of perhaps a signal generator, digital voltmeter, oscilloscope, printer, plotter and computer might be set up. The computer would act as the system controller and would have a program which produced all the commands required to set up the system, carry out a series of measurement, perform calculations on the data and plot or print out results. The bus could be used to set up the signal generator frequency and amplitude and perhaps to step the frequency through a series of values. After each frequency step, the data from the digital voltmeter might be read into the computer memory. If the oscilloscope were a digital storage type it would also be possible to download the contents of the oscilloscope memory to the computer so that the waveform could be analysed. This data might also be sent to the plotter in order to produce a hard copy of the oscilloscope trace. Other data acquired during the experiment having been transferred to computer memory could then be analysed and processed by the computer program and any results output to the printer or the plotter. With such a system each experiment or series of experiments can be programmed into the computer and the instruments then proceed to carry out the measurements automatically.

The IEEE488 bus system

The IEEE488 bus system consists of a total of 16 lines. Of these there are 8 lines (DIO1 – DIO8) which are used for data transfers between instruments connected to the bus system. These lines are bi-directional and are also used to transmit commands to the individual instruments. A further set of five lines are used for bus management signals. The signals on these lines are,

ATN Attention
EOI End or Identify
IFC Interface Clear
REN Remote Enable
SRQ Service Request

Finally three lines are used to provide handshake signals between instruments when data is being passed between them. These lines are NRFD (Not Ready For Data), DAV (Data Available) and NDAC (Not Data Accepted).

All of the lines on the bus are arranged to operate using a wired OR configuration and the signals are arranged so that a low logic level is read as a true or 1 state. Thus the logic convention is the inverse of that on a normal logic system.

The bus connector

The bus interconnections between instruments are implemented using 24 way side contact connectors. The connector fitted to the instrument is always a socket and those at each end of the connecting cable are plugs. When a number of instruments are to be linked together a special piggy-back plug may be used on the connecting cable. This has a socket mounted on top of the plug and allows another cable to be linked in to take the connection on to the next instrument in the chain. Several of these piggyback connectors might be stacked if desired.

The allocation of connector pins to the various bus signals is as follows:

Pin	*Signal*	
1	DIO1	Data 1 (LS)
2	DIO2	Data 2
3	DIO3	Data 3
4	DIO4	Data 4
5	EOI	End or Identify
6	DAV	Data Available
7	NRFD	Not Ready For Data
8	NDAC	Not Data Accepted
9	IFC	Interface Clear
10	SRQ	Service Request
11	ATN	Attention
12	SCRN	Screen
13	DIO5	Data 5
14	DIO6	Data 6
15	DIO7	Data 7
16	DIO8	Data 8 (MS)
17	REN	Remote Enable
18	GND	Ground
19	GND	Ground
20	GND	Ground
21	GND	Ground
22	GND	Ground
23	GND	Ground
24	GND	Ground

The management bus

Let us now look at the functions of the various signals on the five lines that form the management bus. The first of these is the ATN signal which is always issued by the master controller and signifies that a command is about to be placed on the data bus lines. In effect this signal tells all of the other instruments connected to the bus to pay attention. Commands may be either broadcast so that all units on the bus will respond to the command or alternatively they may be addressed to specific instruments and only those instruments respond to the command.

The EOI signal line may be used in two ways. In one mode it is used as an End signal and is activated by an instrument that is transmitting data to the bus. When the block of data has been passed over the bus system the EOI signal is asserted to show that there is no more data to come. In this mode of operation of EOI the ATN signal is not activated. The alternative function for the EOI line is that it may be used by the master controller to request that an instrument requesting service should identify itself. In this case the controller asserts both the ATN and EOI signals.

When an instrument on the bus system has some data ready for transfer to another unit or alternatively requires some data before it can perform its next task the instrument will place a Service Request SRQ signal on the bus. The SRQ lines from all of the instruments are wire ORed by the bus so that any instrument can request service without damaging the SRQ output of other units connected to the bus. When a service request is detected by the controller the first action must be to identify which unit, or units, requested service. If more than one unit has activated the SRQ line some form of priority scheme is normally used to service the requests one at a time until the SRQ signal has been cleared.

The Interface Clear IFC signal is generated by the master controller and has the simple function of resetting the status of all instruments connected to the bus system to some pre-defined default setting.

The Remote Enable REN line is asserted by the controller to allow instruments connected to the bus to be activated. When not remote enabled, the instrument will usually be responding to manual input at its own control panel. When enabled the instrument can be controlled by commands sent to it via the IEEE488 bus.

Commands

The set of commands that can be issued by the controller is as follows:

Command	*Code*	*Function*
LISTEN	001aaaaa	Set aaaaa as listener
TALK	010aaaaa	Set aaaaa as talker
ADDRESS	011aaaaa	Set aaaaa as secondary address
DCL	00010100	Device clear
GET	00001000	Group enable trigger
GTL	00000001	Go to local
LLO	00100001	Local lockout
PPC	00000101	Parallel poll configure
PPD	01110000	Parallel poll disable
PPE	01100000	Parallel poll enable
PPU	00010101	Parallel poll unconfigure
SDC	00000100	Selected device clear
SPD	00011001	Serial poll disable
SPE	00011000	Serial poll enable

Here the code aaaaa is the binary code giving the address of the device to be acted upon.

Listeners and talkers

Instruments connected to the bus may be set to operate as either listeners or talkers. When operating as a talker the instrument is able to output data signals to the data bus lines. To avoid signal conflicts only one talker can be active on the bus system at any particular time. The unit selected as the talker can be switched by a command from the controller so that all of the instruments could become a talker in turn in order to transfer data to the bus.

The alternative mode of operation is as a listener where the instrument is able to receive data from the bus system. Several instruments may act as listeners at the same time since there is no possibility of signal conflict when they are operating in this mode. Thus a single instrument acting as a talker can transmit data to several other instruments which are acting as listeners. Again individual instruments may be selected to act as listeners as desired by sending commands from the master controller. When an instrument is not selected as either a talker or listener it will ignore data on the bus except when the ATN signal is present indicating that a command is being sent from the controller.

A further mode of operation for some instruments on the bus is that of being the bus master controller. Normally this function is provided by a microcomputer which might, for example, be an IBM PC or one of the compatible clones of this series of machines. In most systems only one of the devices in the system is designed to operate as a controller. It is possible to have a system where two or more of the units connected to the bus are capable of operating as a master controller. An example of this might be a system where two or more of the units on the bus are general purpose microcomputers. To allow for this possibility a command is provided which allows the controller to transfer its control function to another unit on the bus. Once the control function has passed to the new unit it can be assigned back to the original unit if the new controller issues the appropriate command. The important thing to remember is that at any time only one unit on the bus can be the master controller.

Device numbers

Up to 31 separate instruments may be connected to the IEEE488 bus and these are identified as device No 0 to device No 30. When the controller wishes to send a command to a specific instrument the device number for that instrument is sent on the data bus and only the specified device will respond. In the instrument itself there will usually be a set of thumbwheel or dual in line switches which can be used to set up the device number to which the instrument will respond. When setting up these switches it is important to ensure that the device number selected is not one that is already being used by another instrument.

If a computer is being used as the master controller and the program being used allocates specific device numbers for certain instruments or measurements, it is important to ensure that the instrument device numbers are set to match the requirements of the program.

When a device is to be set up as a talker the device number code is sent as part of the command word. In this case the upper three bits (DIO8, 7 and 6) are set as 001 and the remaining 5 bits provide a binary number from 0 – 30 which gives the address of the device to be set as a talker. If all five lower bits are set to 1 giving an address of 31 this is treated as a special command called UNTALK which disables the device that is currently set as the talker. This UNTALK command should always be sent before selecting a new device as talker so that only one talker is ever activated at a time.

To set a device as a listener the code 010 is sent on bits 8,7 and 6 of the data bus and again the lower 5 bits are used to convey the device address as a binary code number. The address code 31 is used to provide the special command UNLISTEN which disables all of the devices currently set up as listeners on the bus. If several listeners are to be set up the LISTEN commands including the appropriate device numbers are sent in sequence. To turn off a single listener the UNLISTEN command is sent and then all of the other listeners that were set up will need to be sent a new LISTEN command to set them up again.

For other commands to a specific instrument the address may be sent as part of a sequence of data words making up the complete command. Thus the command word might be sent first and then the address is sent as a pair of ASCII coded numeric characters representing the numbers 00 to 30. The command is then completed either by sending a Carriage Return code (decimal 13) or by sending a signal on the EOI line. An example might be sent as follows:

Data bus	*Function*
00000100	Command (selected device clear)
00110001	ASCII 1
00110101	ASCII 5
00001101	ASCII CR

Here the command sequence starts with a command code (SDC) which is accompanied by a 1 state on the ATN line. This command indicates that the selected device is to be cleared to a default state. The following three data words are sent with ATN set at 0 which indicates that they are data words rather than commands. The first two are ASCII codes for the numbers 1 and 5 which indicate that the device number of the instrument that is to respond is number 15. The final data word is a carriage return code which indicates the end of the command and on receiving this the specified instrument with device number 15 will reset itself. Other units connected to the bus will also detect the SDC command and decode the following address but because the address code does not match the device number to which they are set the command is ignored.

Secondary addresses

The secondary address command is similar in format to the commands for setting listeners or talkers. The code sent has its upper three bits set as 011 and the lower 5 bits provide a secondary address. These secondary addresses are used by the instrument as local commands to perform some function specific to the instrument itself. Two of the secondary address codes are reserved for use as commands for carrying out a parallel poll of device status.

The handshake system

When data is to be transferred across the bus, a handshaking sequence occurs for each data word that is to be transferred. The device placing the data on the bus will first check the NRFD line to see that all devices on the bus are ready to accept data. Once this line has gone to the false condition the data word is placed on the bus and the DAV signal is set true to indicate that data is available. At this point each device will set its NDAC line false as it accepts the data and when all lines are false every listener will have read the data word and new data can be set up. At this point the DAV signal goes false to indicate that the data is being changed. Now the NRFD lines of the various devices will have been set true since they may not yet be ready to accept a new data word and the whole sequence starts again.

Service requests

As mentioned earlier when an instrument has some data ready for transfer or requires some other attention it asserts a signal on the SRQ line. On detecting the signal on the SRQ line the controller must discover which of the instruments on the bus requires attention. There are two methods by which this may be done.

The first approach to finding out which unit has asserted the SRQ signal is to use a serial polling technique. Here the controller sends a command to start a serial polling sequence and on receipt of the command each unit on the bus prepares a status byte indicating the current status of the instrument. The controller then interrogates each instrument in turn and checks its status byte to see if it is the one which requested attention.

The actual sequence of commands on the bus might be as follows:

SPE	Serial poll enable command from controller
TALK 3	Set unit 3 as talker
UNLISTEN	Reset all listeners
MLA	Set controller to listen
data	Status byte from unit 3
SPD	Serial poll disable

Normally when a serial poll scheme is being executed each of the instruments on the bus is allocated a priority level. The first status byte to be checked by the controller will be that from the instrument with the highest priority. If this instrument does not require attention the controller goes on to interrogate the unit with the next lower priority and continues in this fashion until it finds the unit that has requested attention. At this point the controller checks the status byte to see what action is required. The serial poll sequence is then disabled by an SPD (Serial Poll Disable) command. The appropriate actions are then carried out to service the instrument. After this the controller will again check the SRQ line and if it is still activated a new serial poll routine is started.

Parallel poll sequence

Although the serial poll technique can work quite effectively for dealing with service requests it has the disadvantage that when one of the lower priority units requests attention the controller still has to interrogate the status of the higher priority units. There is however an alternative technique which can give an immediate indication of which instrument has requested attention. This is known as the parallel polling technique.

In the parallel poll technique up to 8 instruments may be checked simultaneously. Each of the instruments is allocated one of the eight data bits of the data word as its status bit. When the polling action takes place each instrument sets its allocated status bit and so the status of all eight instruments is transferred as a single word. The controller then examines this word to see which instrument requested attention.

For the parallel poll scheme the first stage is to allocate the various status bits in the status word to the appropriate instruments. This is done by using the command PPC (Parallel Poll Configure) which is followed by a series of device numbers defining the instruments which are allocated to each of the data bits.

To initiate a parallel poll the controller issues a PPE (Parallel Poll Enable) command in response to which the eight instruments configured for parallel polling will place their status bits on the bus to provide a status word. The controller can then disable the poll routine by sending a PPD (Parallel Poll Disable) command and then proceeds to process the service request.

Other commands

In a typical instrumentation system it may be required that several instruments should take readings simultaneously. In this case the GET (Group Execute Trigger) command is sent. Another command is LLO (Local Lock Out) which commands a specified device to disable its local control panel so that it responds only to commands from the IEEE488 bus. The GTL (Go To Local) command has the reverse effect in that it restores control from the instrument's own front panel and disconnects the instrument from the IEEE488 bus. To regain control of the instrument by the bus a REN signal can be issued by the controller.

Index